The Vegetarian Imperative

The Vegetarian Imperative

ANAND M. SAXENA

The Johns Hopkins University Press
BALTIMORE

Printed in the United States of America on acid-free paper
2 4 6 8 9 7 5 3 1

The Johns Hopkins University Press
2715 North Charles Street
Baltimore, Maryland 21218-4363
www.press.jhu.edu

Library of Congress Cataloging-in-Publication Data

Saxena, Anand M.
The vegetarian imperative / Anand M. Saxena.
p. cm.
Includes bibliographical references and index.
ISBN-13: 978-1-4214-0242-0 (hardcover : alk. paper)
ISBN-10: 1-4214-0242-4 (hardcover : alk. paper)
1. Vegetarianism. 2. Sustainable living. I. Title.
RM236.S356 2011
613.2'62—dc22 2011000178

A catalog record for this book is available from the British Library.

Special discounts are available for bulk purchases of this book. For more information, please contact Special Sales at 410-516-6936 or specialsales@press.jhu.edu.

The Johns Hopkins University Press uses environmentally friendly book materials, including recycled text paper that is composed of at least 30 percent post-consumer waste, whenever possible.

CONTENTS

PREFACE

I was born into a vegetarian family that also did not condone eating eggs, although milk and milk products were freely allowed, even appreciated. I remained a vegetarian after leaving home for studies in college, and later on for other pursuits; the only change was that I occasionally indulged in eating an omelet. When I started preparing to come to the United States in 1968 for graduate studies, I was told that being a vegetarian in the States would be very difficult, even impossible. Since I had no strong convictions on the subject, I decided to start eating meat whenever necessary.

My first encounter with meat was in a fancy hotel in Beirut, where I stayed for a day on my way to the United States. I asked the waiter to bring me whatever was the house special for the day, and he brought a dish containing lamb—the favorite food in that part of the world. However, I found that for a lifelong vegetarian, one who had never even come close to a meat preparation, eating meat was extremely difficult. There was an aroma to the dish that was so intolerable to me that I could not bring myself to take more than one bite. After munching bread for a while, I asked the waiter to take the dish away and get me some ice cream. The waiter was somewhat offended in having the whole dish returned and notified the chef, who explained to me how proud he was of his preparation. All I could say in response was that I was not feeling very well and just wanted some rest.

As a graduate student in New York City, I rented part of a house. On some occasions, I helped myself to some meat dishes in the school's cafeteria but could never bring myself to eat them. The house I lived in was owned by a kind elderly lady of Italian extraction who was a retired seamstress. She frequently gave me lectures in support of eating meat, believing that God made these creatures for us. Although she often made a pasta dish for me, on a few occasions she invited me to her house for a spaghetti and meatball dinner. Even though I told her that I had, in principle, no objection to eating meat, I still could not bring myself to eat the meatballs. On most of these occasions, I wrapped the meatballs in tissues and put them in my pocket when the landlady was not looking. Later, I took them to school and threw them in the trash.

The variety of foods available in markets in 1968 was a lot different from what is available these days. Bread meant simply white bread, with specialty breads available only in a few places. There was a small section for yogurt, but it contained only one kind of fruit yogurt made by Dannon. There were fewer vegetables than what one can find in a store today in the same locality, and there was almost a complete absence of Spanish, Chinese, and Indian foods. There was only one Armenian store in Manhattan that kept some Indian food items.

During the academic year 1968-69, I basically survived on rice pilaf, a concoction that I made with rice and a few vegetables, and I ate it with buttermilk. When I needed food between classes, I bought french fries and cottage cheese in the cafeteria. Things changed in the summer of 1969 when I attended a summer program at Brandeis University. I did not have a car and was completely dependent on the school's cafeteria, which closed promptly at 6:30 p.m. Since I had no access to cooking or shopping facilities, and the main courses in the cafeteria were unpalatable to me, I starved for a few days. A nice couple who were attending the same school understood my predicament and decided to help me out. They drove me to places where the food was not so bland and slowly introduced me to meat preparations. By the time summer school had ended, I had become a "nonvegetarian." Some time later I joined Brookhaven National

Laboratory, located in the suburbs about 60 miles east of New York City, to work on my Ph.D. dissertation. While at Brookhaven, I ate whatever was available in the cafeteria because I had neither the time nor the inclination to get involved in cooking food. In most respects, I became an omnivore just like most people around me.

I got married a few years after receiving my degree. My wife had been an omnivore since childhood. She had an intrinsic appreciation for some meat-based preparations and encouraged me to help her make them. For a good many years, we were a typical non-vegetarian family, the only difference being that our consumption of meat and fish was well below the American average because on many days of the week, we ate a diet consisting solely of grains, fruits, vegetables, and milk products. Although I had no particular aversion to meat dishes, I did not develop a great liking for them.

I was mildly interested in a few books that came out around 1990, particularly those by Frances Moore Lappé, which showed that a vegetarian diet is better for our overpopulated planet. However, my curiosity was particularly aroused when I read some scientific journal articles that assessed the adverse effect on the environment of eating meat, and I slowly began to compile these papers. In 1997, my daughter started attending Barnard College of Columbia University and announced that she had become a vegetarian. In response to demand from a number of young women, the cafeteria in her school had started a section for vegetarians. To me, it was interesting to learn that there were so many young persons interested in vegetarianism. Soon thereafter, the wealth of information on the subject convinced me to go back to my roots and become a vegetarian.

As I learned more about the effects of vegetarian and omnivorous diets, I was amazed at the amount of credible scientific information available on the subject. There were basically two types of analyses and research papers that interested me. The first of these concerned the effects on the ecosystem of eating meat and the second the beneficial effects of plant foods on human health. Since the ecosystem is vast and complex, the effects of a meat-based diet on the ecosystem are usually recorded and cataloged by national or international organizations. A very large number of papers in medical, health, and nutrition-related journals have shown the benefits of diets based on grains, vegetables, fruits, and nuts. These facts have become so firmly established that they are frequently mentioned in health-related discussions in the media. As my knowledge increased, I was startled by the range of arguments in support of a plant-based diet.

The time line of environmental changes that were taking place demanded a sense of urgency. I decided to take my ideas to discussion forums in libraries and other places. I was surprised to find that most people were completely unaware of the details of our food industry and its impact on the planetary resources and the ecosystem. I also learned that changing food preferences is not easy, even for people who seem to agree with the basic premise behind the desired change. Out of the numerous arguments in favor of changing our dietary pattern, if I had to pick one aspect that overpowers the others, it is that we want to leave to our children a sustainable future. I am sure that this is true for most of us—few people want to change the balance of nature in such a way that our children have limited choices or may even face deprivations. With this as the central argument, the case for vegetarianism can be augmented with other aspects of the importance of our food choices.

There are numerous publications from national and international organizations that forecast the demand for various kinds of foods, including meats, during the coming decades by extrapolating from the current trend. Since these projections always indicate that the planet has to produce a lot more food than at present—and it is already difficult to produce the amount of food we currently have—they consider all kinds of ways to meet the demand, including methods that are beyond the reach of science at present and those that would cause excessive harm to the ecosystem. I am bothered by the inflexibility that is built into these projections on the demand side of the equation. Human beings are intelligent, thinking people. We should not bury our heads in the sand like ostriches and wait for things to happen. Instead of considering the demand as fixed and stretching our imaginations to meet the supply, why not educate people to change the demand side of the equation? It may not be easy, because it involves convincing a whole lot of people to make different choices. However, it is much better than the alternatives. I believe that it can be done. We need to disseminate persuasive information, and that is why I wrote this book.

ACKNOWLEDGMENTS

I am indebted to my wife, Mala, who provided invaluable support and encouragement that enabled me to complete this work. Our daughter Nalini provided me with very useful reading material relevant to the ideas that I was developing and provided some insight into the data plots and figures. Our son Anuj read a few chapters and provided very helpful comments to improve their presentation.

I am grateful to the acquisitions and editorial staff of the Johns Hopkins University Press, in particular Jacqueline Wehmueller and Deborah Bors, for expertly editing the manuscript during various phases of its preparation and also for numerous suggestions that make the text more cogent.

This work could not have been possible without the extensive information and databases maintained by the United States Department of Agriculture, Environmental Protection Agency, Geological Survey, and Centers for Disease Control and Prevention; the World Resources Institute and the American Institute for Cancer Research; and various agencies of the United Nations, including the Food and Agriculture Organization and the Environment Programme.

The Vegetarian Imperative

INTRODUCTION

We have not learned to take food seriously. A person living in the developed world seems to have an endless variety of food choices, from breakfast cereals and lunch meats to snack items. We munch all through the day. However, there are vitally important reasons for taking food seriously. Put simply, if we do not change our ways, we humans will eventually run out of food because our planet will not be able to produce what we need. We need to take food seriously to save the environment (or at least not despoil it to the extent that our children have to suffer deprivations), to provide sufficient food for people living right now and in the future, and to protect our own health.

Most of us will be surprised to learn the amount of food that we consume on an annual basis. The food eaten by an average American per year includes 214 pounds of meat (red meat, chicken, and fish), 32 pounds of eggs, 600 pounds of milk or its equivalent in dairy products, 644 pounds of fruits and vegetables, 67 pounds of potatoes, more than 150 pounds of grains, 46 gallons of carbonated beverages, and 26 gallons of alcoholic beverages.[1] This list does not include items such as coffee, tea, and chocolate. Consumption in the United States is not representative of consumption in the developed world, but other industrialized societies are only 10 to 20 percent behind us. The amount of food consumed by people in the developing world depends on the economic status of the individuals.

The number of people inhabiting the earth is approaching seven

Per capita meat consumption, in pounds.

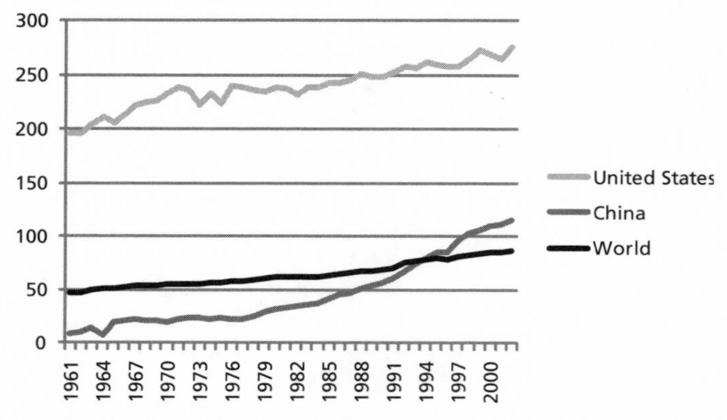

Source: World Resources Institute, http://earthtrends.wri.org/.

billion. The food requirements of this population force us to transport food from all corners of the world, sometimes to meet basic necessities and at other times to satisfy our desire for a particular kind of product. There is no part of the planet that has not been explored for edible material. The global demand for food for the present population is already very large, and it is on an upward spiral, increasing with each passing year. There are two main reasons for this situation. First, the population of the world is growing by 76 million every year and is projected to reach 9.1 billion by the year 2050. Second, and at least as important, a nutritional transition is taking place in most regions of the world, with a greater demand for foods of animal origin—meat, milk products, and eggs. The consumption of these items in the developed world, while already very high, is still increasing at a slow rate, while in the developing world the demand for animal-based foods is increasing rapidly.

There are two main drivers of this nutritional transition. During the last few decades, the per capita income has increased in much of the world. In general, it has been observed that prosperity almost always induces people to eat more energy-rich foods. During the 40-year period from 1963 to 2003, the per capita consumption of calories from meat in the developing world increased 119 percent. In China, the per capita consumption of meat during this period increased 349 percent.[2] Since China is the most populous country in the world, this change is having a large impact on the global demand for food. The upward trend in the demand for foods of animal origin is accelerating in all developing countries.

A number of factors have contributed to a nutritional transition in favor of foods of animal origin. Our metabolism, which developed in the Paleolithic era, when food was not easily available at all times, creates an intrinsic desire for energy-rich foods. This tendency is compounded by the widely held (and incorrect) belief that a diet containing a lot of meat, milk, and eggs is better for health than one containing mostly grains, fruits, and vegetables. In addition, mass media like television and the Internet have made people across the world familiar with the lifestyles of those in the developed world and have created a clamor for emulation, at least among the emerging middle class in the developing countries.

Other factors that contribute to this nutritional transition include urbanization, marketing by the food industry, and trade liberalization. In the beginning of the twentieth century, only 10 percent of the population lived in cities; the proportion is now close to 50 percent and is increasing rapidly in the developed world as more and more people migrate to cities in search of jobs and a better life. People living in cities are more influenced by the mass media. Cities also provide a distribution infrastructure for large, often multinational, corporations. Liberalization of trade, which has become a catchphrase these days, affects the availability of certain foods by removing barriers to foreign investment in food distribution. Because of these developments, the multinational fast food industry has outlets in many cities in the developing world; these restaurants then become agents of change in favor of fast food.

In the final analysis, all human activities are performed with the expenditure of energy, and the activities' viability depends on the relative consumption and availability of energy. The primary source of energy on earth is the sun; the only terrestrial source of energy—nuclear energy—makes a very small contribution. Contemporary lifestyles in developed countries—with frequent long-range travel, air-conditioned homes, and access to factories that produce objects to support this lifestyle—consume much more energy than can currently be harvested from the sun, the wind, and water, either naturally (through photosynthesis by plants) or artificially (through solar cells, windmills, hydroelectric plants, and other devices). The additional energy needed for this contemporary lifestyle is provided by fossil fuels. More than 85 percent of the energy used in the United States is derived from fossil fuels,[3] making us highly dependent on this source of energy. The energy derived from the three kinds of fossil fuels—oil, natural gas, and coal—is also solar energy, but it was harvested by an enormous number of microorganisms millions of years ago and is now buried underground. In a way, fossil energy represents our inheritance from previous ages: vast but increasingly limited reserves. The use of energy from any of these fuels is not benign for the environment. At the very minimum, burning of fossil fuels releases the greenhouse gas carbon dioxide into the atmosphere, but the combustion process also gives rise to many other harmful gases and airborne particles. The factories that use fossil fuels to synthesize various products invariably pollute the air and water in those regions.

While solar energy can be harvested and put to use with a simple device like a solar cell, the production of food for human consumption requires many other inputs and suitable conditions in addition to sunlight. Deserts with plenty of sunlight cannot produce food for us; even water, the essential first requirement, by itself cannot produce harvests of grains or other edible products in sufficient amounts to feed the present population. Besides sunlight and water, the production of food on a large scale requires suitable soil, the proper range of temperatures when the crops are growing, and numerous beneficial microorganisms.

Eating the primary agricultural products—grains, fruits, vegetables, and nuts—is called eating closer to the sun because there are no intermediary steps. Feeding agricultural products to farm animals and then consuming animals as food is a secondary process, with a large concomitant loss of energy; thus, producing these foods increases the burden on the ecosphere. Our choice of food is important because it determines the quantity of primary agricultural products used to fuel our activities.

Two resources that are essential for large-scale agricultural activities are water and suitable land. Agriculture requires vast quantities of water; almost 70 percent of water used in all anthropogenic activities is used to support agriculture. Global water resources are already under stress at the current population levels, which will only increase in the coming decades. Many lakes and rivers are vanishing, and the quality of those that remain is deteriorating. Groundwater supplies are under pressure from overuse and pollution. Almost all the land that is suitable for farming is already being cultivated; any addition can only be made by razing rainforests, with serious consequences for the global climate and precipitation pattern.

Modern agriculture has become highly dependent on fossil fuels. Nitrogen fertilizer in a form that plants can use is produced from natural gas and some coal and accounts for 30 to 50 percent of total energy use in commercial agriculture;[4] potassium, phosphorus and some other nutrients in suitable forms are usually added to the nitrogen fertilizers. Other ways in which large amounts of fossil fuels are used in agriculture include the operation of field machinery, transportation, irrigation, and the production of pesticides.[5] The commonly used pesticides include herbicides, insecticides, and fungicides, representing roughly 50 percent, 30 percent, and 10 percent of total pesticide usage, respectively.[6] Because of the large quantities of chemicals, including fertilizers, used in contemporary agriculture—primarily because modern varieties of corn, wheat, and rice grown nonorganically depend on regular application of these chemicals—the amount of fossil energy used in growing grains may be more than the energy content of the grains. Hence, it may be a truism to say that we are literally eating fossil fuels. As the demand to produce food increases, agriculture becomes even more dependent on fossil fuels as the amount of applied fertilizers increases. However, we have reached the stage of diminishing returns, because additional use of fertilizer does not produce the same increase in output as earlier, but it does add to environmental degradation, including degradation of the soil due to the accumulation of some chemicals.

Intensive nonorganic agriculture does not allow the land time to regenerate its lost capacity. The use of chemicals degrades farmlands. Nonorganic agriculture causes a loss of topsoil and kills beneficial microorganisms. By depleting planetary resources and degrading the environment, the amount of food that we force the land to produce, which greatly depends on the choice of food that we eat, affects the land's capacity to continue to provide food to us and to future generations. Animal foods require much more energy to produce than plant foods do. This is particularly important these days because the vast amount of food needed to feed the human population is already approaching the limit of planetary capacity. If all factors are taken into consideration, the adverse impact on the ecosystem of a diet rich in animal products is greater than that of any other human activity.

The demand for foods of animal origin causes the livestock industry to raise and slaughter a very large number of animals. According to the U.S. Department of Agriculture (USDA), the live weight of farm animals at any time is about five times that of the human population in this country.[7] The ratio for the entire world is not quite so skewed because of a lower consumption of meat in many countries, but the farm animals still outweigh the human population of close to seven billion by at least a factor of two. Since these animals only live during the time when they are rapidly putting on weight, their burden on the ecosystem is even greater than indicated by their weight. In addition to consuming resources, farm animals are major contributors to climate change, both directly, through emissions of greenhouse gases, and indirectly, through the use of fossil fuels in the agriculture that provides their food. Climate change may already be rearranging rainfall and glacial patterns, making life in arid areas increasingly untenable and intensifying floods in the already wet, and more populous, regions. The availability of food will be adversely affected by extreme weather and ecological stress caused by global warming. Together, these and other forces will challenge the capacity of the world's food production system.

There are yet other reasons to take the choice of food seriously. Although residents of the developed world are not affected by it to a significant extent, the world's food situation is at best precarious. The amount of food that is produced these days is not sufficient to meet the needs of the present population, with the result that more than a billion people remain hungry or are malnourished for want of food.[8] Because of the loss of energy and nutrition that oc-

World meat production, in millions of tons.

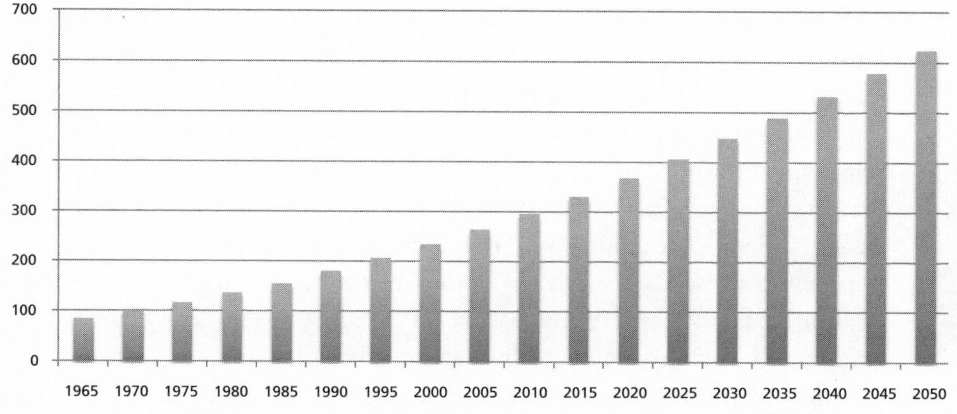

Source: U.N. Food and Agriculture Organization, *World Agriculture: Towards 2030/2050,* www.fao.org /fileadmin/user_upload/esag/docs/Interim_report_AT2050web.pdf.

curs when converting agricultural products to animal-based foods, eating such foods is wasteful of the planetary budget. As the demand for food increases during the coming years, the situation will become progressively worse. Taking into account the increasing population and changing food preferences, the global output of agricultural products has to double by 2050. This calculation does not even address the challenge of reducing hunger in the world. Attempts to increase the production of food by razing forests or excessive use of fertilizers will lead to rapid deterioration of the environment and severe consequences for the global climate.

A major shift in food preferences would go a long way toward alleviating hunger in the world, both now and in the future. An average person in the United States has a daily intake of 3700 kilocalories (kcal, often called calories) per day, about one-third of which comes from animal-based foods. A change in food preferences in favor of primary agricultural products in both developed and emerging economies would release a substantial amount of food for direct human consumption and also curb the increase in the global requirement for agricultural products consumed by animals.

The proclivity of humans to eat energy-rich foods, if they can find and afford them, has other consequences as well. People in developed countries, and an increasing number in emerging economies, suffer from noncommunicable chronic diseases, including heart disease, diabetes, cancer, and neurodegenerative diseases. Twenty-four million Americans (11 percent of the noninstitutionalized adult population) have heart disease, which is the number one cause of death in the country. Cancer and stroke are numbers two and three in terms of the toll on human lives and afflict 15.8 million and 5.6 million persons, respectively. Ten percent of adults suffer from diabetes.[9] Excessive consumption of high-calorie and more energy-dense food increases the incidence of obesity—a global epidemic these days—which causes morbidity by itself and is a causative factor for chronic diseases. These diseases not only hasten mortality but also decrease human productivity and degrade the quality of life for many years. The important thing about chronic diseases is that many of them are lifestyle diseases; their incidence can be reduced by as much as 30 to 40 percent with the proper combination of diet and exercise. During the last few years, there has been a paradigm shift among nutritionists and health professionals in favor of vegetarianism.[10] A properly chosen vegetarian diet, even without other lifestyle changes, can contribute to lowering mortality as well as morbidity.

The prospect of ever-growing demands being placed on increasingly degraded ecosystems seriously endangers even the status quo—the current state of nutrition and hunger in the world. Events are moving at such a rapid pace that serious consequences of our lack of concern for the environment may become visible within two or three decades, within the lifetime of our children, if not our own lifetime. Relying on the market, government, or science to solve these problems that are looming on the horizon is not a judicious approach. Some scientific developments, government regulation, or market corrections may provide a temporary reprieve, but there is going to be an inevitable collision between the increasing demands of the growing population and the capacity of the ecosystem to sustain injuries while still being forced to meet greater and greater human needs. Although reducing the consumption of animal products will not solve all the problems that have been caused by various factors within the last few decades, it is one step that can be easily taken and will have a positive effect more quickly than other possible options. For this reason, drastically reducing the consumption of animal-based foods has to be a necessary first step in any environmental movement to preserve the planetary resources for the coming generations. While we all have to eat, we can educate ourselves and make better food choices to reduce the harmful consequences of our diet.

1 * FARMS

At any given time, billions of animals are being raised around the world: one and a half billion cattle, close to a billion swine, close to two billion sheep, goats, and buffalos, and 17 trillion chickens.[1] But not all of these animals are being raised in the same way. For example, in Africa, nomads travel with their animals to wherever enough water and vegetation can be found. Many animals around the world are raised in urban and semiurban settings, where they spend little time in pasture. A commonly held ideal is of livestock being raised on traditional family farms where they happily spend their days doing what comes naturally to them. Cows munching on grass in a meadow, chickens pecking to find a worm or a speck of grain, pigs frolicking with each other—these are the symbols of the pastoral setting. This is the image evoked in the poem "The Cow," by Robert Louis Stevenson:

> The friendly cow all red and white,
> I love with all my heart:
> She gives me cream with all her might,
> To eat with apple-tart.
>
> She wanders lowing here and there,
> And yet she cannot stray,
> All in the pleasant open air,
> The pleasant light of day;

And blown by all the winds that pass
And wet with all the showers,
She walks among the meadow grass
And eats the meadow flowers.[2]

Naturally we want to associate the meat on our plate with health and good life; to do this, however, we need to believe that the animals we eat are being raised in a bucolic setting. Unfortunately, this image of farm animals leading peaceful lives is far from reality—or, more accurately, it is a reality only to a limited extent. For a variety of reasons, small family-style farms whose animals are maintained with care by a farmer are increasingly being replaced by industrial facilities known as Concentrated Animal Feeding Operations (CAFOs). These operations raise tens of thousands of animals in factory-style settings.

Raising animals in large, centralized facilities, pejoratively called animal factories, is the dominant mode of production in Europe and North America and has gained considerable acceptance in Eastern Europe and Central America. More recently, CAFOs have become the most common way of raising farm animals in most Asian countries as well, including Malaysia, the Philippines, Thailand, and China. According to the U.S. Environmental Protection Agency (EPA), a facility that houses more than 1,000 cattle, 2,500 swine, or 125,000 chickens is classified as a large CAFO. Many industrial operations maintain much larger numbers of animals. Worldwide, such operations generate 74 percent of the world's poultry products, 50 percent of all pork, 43 percent of beef, and 68 percent of eggs.[3] In developed countries, almost all chicken and swine are raised in CAFOs, and the proportion of cattle in such facilities is greater than 70 percent.

The number of animal farms in the United States is still fairly large, but the contribution of smaller farms to the production of foods of animal origin is very small. Nearly half of the country's livestock of all types are produced on just 5 percent of the farms. In Texas, one of the four major states (the others are Iowa, Kansas, and Nebraska) that together supply 52 percent of the red meat in

Table 1.1. Farm animals (excluding poultry) in various regions, in millions, 2006

Region	*Cattle*	*Swine*	*Sheep/Goats*	*Buffalo/Camels*
North America	112	76	10	0
Central America	50	21	24	0
South America	344	53	96	1
Asia (excluding Middle East)	430	616	815	172
Middle East/North Africa	33	0.7	233	7
Sub-Saharan Africa	223	24	371	9
Europe	128	192	155	0.2
Oceania	39	3	141	0
Total	1,383	990	1,939	190

Source: World Resources Institute, http://earthtrends.wri.org/.

Table 1.2. Farm animal nomenclature

Cows: General term used for cattle of all kinds
- Heifer—female that has not given birth to a calf
- Steer—young castrated male
- Ox—mature castrated male
- Bull—uncastrated male
- Calf—young cow

Chickens: *Gallus gallus domesticus,* the most common bird raised for food
- Broiler—raised for its meat
- Layer—female raised for laying eggs
- Pullet—young female

Swine: General term used for this family of animals
- Pig—young swine less than 120 pounds in weight
- Piglet—young pig
- Hog—swine more than 120 pounds in weight
- Boar—male that has not been castrated
- Sow—female raised for delivering piglets
- Runt—weakest young in a litter, often killed

the United States, nearly half of the permitted beef cattle CAFOs hold more than 16,000 animals.[4] This trend toward confining large numbers of livestock in factories is increasing. Small operations are disappearing at the rate of 1 to 2 percent per year while the bigger ones are increasing in both size and capacity. Many of the smaller farms that still exist take care of animals for only a short time before supplying them to industrial operations; thus these small farms are not truly independent, and the animals they produce are not raised on small farms all their lives.

The lives of animals raised in large-scale industrial operations are very different from the lives of animals on family farms; a major reason is that these large operations can only thrive due to technological advances that change the "lives" of "livestock." The feed of the animals, the lighting, and other factors in their environment have been engineered with the sole objective of increasing output—more eggs per hen, more milk from each cow, and steer,

Table 1.3. Livestock in the United States, 2007

Cattle and calves, excluding dairy	97 million
Number slaughtered each year	36 million
Average weight	1,245 pounds
Milk cows	9 million
Heifers kept for replacement	4.5 million
Milk production per milk cow	20,267 pounds
Broilers slaughtered	9 billion
Turkeys slaughtered	272 million
Laying hens	344 million
Total egg production	91 billion
Hogs and pigs	67 million
Pigs slaughtered	109 million

Source: U.S. Department of Agriculture, *2007 U.S. Animal Health Report* and *Agriculture Information Bulletin 803*.

Table 1.4. U.S. states with more than 4,000 CAFOs

State	*Number of CAFOs*
Iowa	10,718
Texas	7,192
Minnesota	5,480
North Carolina	5,039
Wisconsin	4,839
Kansas	4,695
Arkansas	4,235
California	4,055

Source: U.S. Department of Agriculture, Economic Research Service, www.ers.usda.gov/.

pigs, and chickens that reach market weight in a fraction of the time they used to. Genetic selection has produced breeds of animals that are more productive from a commercial point of view.

One of the important differences between animals raised in traditional settings and those living in CAFOs is what they are fed. While the bulk of the feed on family farms consists of hay and agricultural waste products occasionally supplemented with other items, the feed of animals in industrial operations is carefully formulated to achieve rapid body growth. Its main components are corn, soybeans, and grains. Beef tallow and the blood of cows and chickens is usually included in feed pellets because these components have a high energy content. Eating these materials goes against the animals' nature, but the materials are sanitized and processed so the animals will eat them. Notably, cow meat is not included in the feed of cattle because mad cow disease (discussed below) spreads when the brain or spinal cord of a diseased cow is fed to other cattle. When cattle are afflicted, the carcasses must be destroyed because the disease can spread to humans who eat the cattle.

Consider other conditions in the CAFOs: animals are forced to live in highly cramped conditions, where the space available to each animal is the minimum necessary for its survival. They do not even have room to move. When animals cannot move, a greater proportion of the energy in the feed goes into the growth of their bodies instead of consumption for muscular activities. Extremely crowded conditions, lack of exercise, an inability to perform even simple bodily functions (such as scratching, grooming, swatting insects, shifting weight from limb to limb to relieve stress), and an unnatural energy-rich diet make the animals highly susceptible to sickness and disease. To keep them healthy and to make them grow quickly, a cocktail of drugs, growth hormones, and other chemicals are given to them. Since animals have to remain alive for only a short part of what would be their natural lives, the influence of these living conditions on their overall health is deemed unimportant by the corporations that raise them.

Housing animals in small structures minimizes the companies' costs, because heating, lighting, pumps, and other electrical appliances (such as those used to move the feed and manure) can be kept to a minimum. These large-scale industrial operations produce vast quantities of meat at low cost through economies of scale, mechanization, and cheap labor. A CAFO operates very much like an assembly line in other kinds of factories; the workers often are aware of only their small part in the process. They know little about the overall operation, they perform narrowly defined, menial tasks, and they are easy to replace. Machines are used in various phases of these operations, making it possible for a small staff to service a large number of animals.

CAFOs are sometimes said to represent the landless farming of animals. While the physical area occupied by these operations is very small considering the number of animals they produce, these facilities depend on the resources of places far away, even across continents, and hence their impact is much larger than their footprint. Transportation companies play a major role in the modern livestock industry because feed, animals, carcasses, and supplies must be moved over long distances. These factories also require copious amounts of water, fuel, and cheap labor to operate in a profitable manner.

What explains the growth of large CAFOs? Think of the Walmart model: large corporations can control industries by producing cheaper products through economies of scale, including cheap labor. The subsidiaries and affiliates of large companies allow CAFOs to control all phases of the food production and distribution system, including merchandising grains, producing animal feed and fertilizers, transporting commodities, processing beef, pork, and poultry, and, finally, delivering animal-based foods to warehouses. Today, a few large corporations, such as Cargill, Archer Daniels Midland, Smithfield, and Tyson, dominate the production of meat, poultry, and dairy products. Small-scale farmers or ranchers are paid as contractors to perform well-defined tasks that cannot be achieved in a centralized facility, such as short-term grazing of animals.

By removing livestock from their natural surroundings and highly regulating all phases of their lives from birth to the dismemberment of their bodies, industrial farming has obliterated the relationship between humans and farm animals. Unlike livestock raised on a traditional farm, in these animal factories, individual animals have no identity. Nor does a single animal have much value, because the CAFOs maintain a stock of mature females that they artificially inseminate to replace any animal that is not performing at the desired level.

The scale of these operations, their density of animals, their insistence on bringing the products to the market in the shortest possible time, and various other cost-cutting measures vastly increase the impact of these operations not only on their own long-term environmental sustainability but also on the production of all kinds of foods. The sheer number of animals that are being raised today has created significant environmental problems, as we will see. The industrial mode of production that houses tens or hundreds of thousands of animals in confined spaces and depends on the resources of other large regions of the land increases their environmental impact in many ways. Examining the lives of these animals explains why.

CATTLE

Different breeds of cattle are raised for meat as opposed to milk. Cattle suitable for the production of beef are able to put on weight in a short time, and they have a large amount of intramuscular fat. The USDA requires beef to be labeled according to the amount of fat in the carcass. A marbling score is assigned to the carcass based on the fat content found between the 12th and 13th rib. Carcasses with a greater amount of fat get a better marbling score and command a higher price in the market. The highest grades, in descending order, are prime, choice, and select; grades lower than select are rarely stamped on meats. The breeds of cattle that produce meat with the highest marbling score are Hereford, Angus, and Texas Longhorn, but the amount of fat also depends on the feed given to them. A rich feed consisting of grains and soybeans promotes the development of fat in the body.

In contrast, the only criterion that determines the suitability of a breed for the dairy industry is the amount of milk that a cow can produce under optimal conditions. Contemporary breeds of cows can produce 1,200 to 4,000 gallons of milk per year. A breed suitable for the production of milk cannot be used for beef, because it will not put on enough flesh of the desired kind in a short period. Similarly, beef cows make poor milk producers.

In industrial establishments, beef cattle are raised in a very different way from dairy cows. A calf that is raised for producing beef, usually about 80 pounds at birth, is sent to a rancher soon after weaning to spend the first six to eight months of life on pasture. Ranchers all over the country work as contractors for large corporations; they take the cattle to graze on public or private lands. Most male cattle are castrated early in life because the meat of a steer is more tender and has more fat than the meat of a bull. A stamp from the USDA will clearly distinguish the meat of a bull from that of a steer, inevitably lowering its price.

Most beef cattle receive implants of hormones to boost their growth rate and build up their muscles. Some of these hormones are similar to those taken by athletes for enhancing their performance, while others shut down the reproductive cycle so that a greater portion of the ingested energy is directed toward the growth of their bodies. A large fraction of these hormones remains in the body of the animal, and some of it is still there when the food reaches the consumer. A significant amount of the hormones also passes through the animal's body in the form of waste material and eventually ends up in the environment.

When they are six to eight months old, cattle are sent to feedlots in a CAFO for "finishing"—that is, for making them heavy and meaty enough for slaughter. They are either kept in metal cages or packed together in chambers at such a high stocking density that they bump into each other with every move. Cattle stand on slatted metal floors and their feces and urine are flushed away beneath them. This liquefied manure is pumped into open-air ponds called lagoons. These ponds are often very large and may contain millions of gallons of liquefied waste.

The animals typically spend ten to fourteen months in these feedlots, where their living conditions and feed are regulated to increase their weight in the shortest possible time. Their diet is rich in nutrients and total calories and is designed so their weight increases at the rate of roughly three pounds each day. By comparison, a steer that is fed grass and forage gains less than a pound per day at a corresponding age. In addition to corn, soybeans, and grains, industrial feed consists of alfalfa, tubers, and pulses. Liquefied fats, vitamins, and antibiotics are also often added to the feed.

As noted above, the natural diet of a cow consists of hay and forage; such a diet is healthy for the animal. In contrast, an energy-rich diet causes the animals to have serious physiological problems, such as liver abscesses and stomach lesions, which means that the animals must receive drugs continually to prevent and treat these problems. As a result of breeding advances and rich feed, beef cattle raised in these facilities reach the market weight of 1,100 to 1,200 pounds before they are two years old. Grass-fed cattle are not ready for slaughter until the age of three or four years, and even then they are not as heavy as animals raised in feedlots.

Mad cow disease (mentioned above) causes microscopic holes to develop in an animal's brain. Once infected, the animal quickly suffers the consequences of the disease. It deteriorates physically and mentally, becomes debilitated, and dies within a few months. Mad cow disease is a cause of great concern because it can be transmitted to other cattle or even humans if they eat the brain or spinal cord of an infected animal. The human equivalent of mad cow disease is known as BSE (bovine spongiform encephalopathy) or CJD (Creutzfeldt-Jakob disease). According to the National CJD Surveillance Unit of the University of Edinburgh, more than two hundred persons in Western countries have succumbed to this disease, the greatest number in the United Kingdom and France.[5] The fear that BSE may spread from one infected animal to others in the herd, and may then infect a large number of people, has caused the U.S. government to ban the flesh (muscle and cartilage) of cows in their feed, although it may be included in the feed of animals that do not get mad cow disease, such as chickens and swine. However, this ban still allows the fat of cattle to be mixed into the feed, which provides an inexpensive source of calories for fattening steer.

Animals such as cattle, goats, sheep, and camels can easily digest fibrous material because they have a highly specialized digestive organ known as the rumen, which is inhabited by billions of microorganisms—bacteria, fungi, and protozoa. When ruminant animals (or "ruminants") eat, the food first goes to the rumen, where microorganisms cause it to ferment, and the fermentation process in turn converts the roughage into microbial proteins and short-chain fatty acids. This semidigested material moves into the true stomach and provides nutrition which sustains the animal, helps it grow (if it is still growing), and, in a dairy cow, allows it to produce milk. Ruminants can digest fibrous grasses, waste materials from fruits and vegetables, residues from seeds from which oil has been extracted, and cellulose from discarded plant products. Their capacity to convert useless material into food for humans has made cattle valuable animals for thousands of years.

When ruminants are fed fibrous material and roughage, their rumens maintain a neutral pH (acid/base level), which is the proper environment for the microorganisms residing there. When cattle are fed grains, however, the rumen becomes severely acidic due to excessive growth of starch-fermenting bacteria that produce a foam, causing the rumen to expand. As noted above, animals living in CAFOs are fed mostly corn, not roughage; as a result, they sometimes develop "feedlot bloat," which leads to inflammation, ulceration, and scarring of the inner lining of the stomach and may even kill the animal by compressing the lungs.[6] After slaughter, about 13 percent of feedlot cattle's livers are condemned because of bacterial abscesses.[7]

Feedlot cattle are given a variety of antibiotics, including penicillin, tetracycline, and erythromycin—the same drugs that are used to treat human diseases. One of the reasons for the routine administration of antibiotics is to prevent epidemics caused by pathogens that may result in substantial loss to the herd. In addition, the operators of CAFOs discovered early on that antibiotics given in low doses, even when there is no sign of disease, make livestock gain weight at a faster rate. Scientists do not fully understand the reason for this phenomenon. In CAFO settings, antibiotics and other drugs are given to the animals to keep them alive and growing while eating a diet that is efficient and inexpensive but unhealthful for them. The primary reason for administering drugs to the animals is to maintain the productivity of the farm.

The living conditions of dairy cows are significantly different from those of beef cattle. Through selective breeding and strict control of diet and environment, a modern dairy cow is unnaturally made to produce an enormous amount of milk, often up to 10 gallons each day. During the last four or five decades, milk production per cow has increased by roughly a factor of four. Thousands of cows are typically kept in a facility, either in indoor stalls or on outdoor drylots. During the milk-producing phase, they spend 10 months at a time, without interruption, living in concrete stalls or on metal floors. These stalls are usually cramped and filthy, since they are not cleaned very often and dairy cows produce an enormous amount of waste. These animals spend their entire productive lives on hard floors with no activity except eating a very rich and unnatural diet and producing milk. It is in the industry's interest to keep the cows pregnant or lactating at all times, so any cow whose milk production declines must be artificially inseminated.

A cow's milk production naturally declines about 70 days after the birth of a calf, but its production can be boosted by an additional 15 percent if the cow is given an injection of bovine growth hormone (BGH, also known as bovine somatotropin, or BST). BST increases both the duration of milk production and the quantity of milk, but it also increases the chance that a cow will develop a condition called mastitis. Mastitis is a painful inflammation of the mammary glands that lacerates the teats and udders. It affects almost 25 percent of dairy cows in CAFOs and is largely a result of excessive milk production. It is aggravated by unhygienic living conditions, like those in CAFOs, where cows recline in unclean stalls. In extreme cases, it causes the blood of the cow to ooze out and mix with the milk.

Cows are given antibiotics to cure diseases such as mastitis, and many of them are also given tranquilizers and sedatives to calm their unexercised muscles and nervous systems so they can be handled without excessive force. Small amounts of these chemicals may be present in the milk. Drugs and hormones not assimilated into their bodies and milk end up in the lagoons that hold their waste products.

In natural settings, a cow will live for 20 to 25 years, but in a typical industrial operation she will be lucky to live for 5 years, because she will be slaughtered as soon as her productivity begins to decline. Cows that are given growth hormones are usually culled during the third lactation because their bodies wear out early; some of this stress is caused by higher milk production. The bodies of dairy cows are weak as a result of their continuous milk production, so their muscles are flaccid and their meat is of poor quality and is usually turned into processed beef or inexpensive hamburgers.

Since the male offspring of dairy cows are not suitable for the production of beef, the common practice has been to raise them for veal. During the 14 to 16 weeks of their brief lives, veal calves spend most of their time in complete darkness in very small wooden crates, where they cannot even turn around. The producers feed the animals a diet deficient in iron and restrict the young calves' movements to inhibit their muscle growth and produce flesh that is tender and pale, both characteristics that are considered appealing in gourmet veal. The crates are kept dark to keep the calves quiet and to reduce their restlessness and boredom from standing in bare wooden crates. By the time the calves are killed, they are often so anemic, sick, and weak that they are near death anyway. A few states in the United States have now banned the confinement of veal calves in crates.

SWINE

From the point of view of livestock farming, the main differences between cows and pigs are that pigs are only raised for meat and, not being ruminants, they cannot digest fibrous materials like grasses and hay. Pigs are omnivores; they will eat just about anything that has nutritional value. There are two main types of industrial operations for raising pigs—one kind keeps only sows in large numbers to be used as factories for making piglets, while others raise pigs until they are ready for slaughter.

Mature sows are usually kept in metallic "gestation crates" from the time they are old enough to be impregnated to the end of their lives. These crates have bare concrete floors; the space is so tight that sows can only rest on their hind legs and are unable to turn around. They typically give birth to nine or ten piglets per litter, and then they are artificially impregnated and the cycle begins again. This process continues until their bodies wear out at about three years of age after delivering three to five litters, at which time they are slaughtered. Against all their instincts, sows must give birth, nurse the piglets, eat, sleep, and defecate, all in the same cramped space. Because of the obvious cruelty resulting from their use, gestation crates have been outlawed in Sweden and the United Kingdom, and some other European countries are considering phasing them out. A few states in the United States have now banned the use of these crates. Smithfield, the nation's largest pork producer, declared in 2007 that it would stop using them in 10 years.

After being separated from their mothers, piglets are kept in pens or cages and while still very young are subjected to a number of surgical procedures: their tails are cut off, some of their teeth are removed, and male piglets are castrated because the meat of boars is considered to have unpleasant odors. When piglets are about two months old and weigh about 50 pounds, they are shipped to a "finishing" facility where they are kept in giant, warehouse-like sheds for four months until they reach an ideal slaughter weight of 250 pounds. These finishing facilities are huge operations that may house tens of thousands of animals; one journalist has observed that they "resemble prison more than farms."[8] The pigs' feed at different stages of life is carefully chosen to increase their weight at a rapid rate so that they are ready for slaughter as soon as possible. Just as with cattle, pigs are routinely given antibiotics and hormones to prevent the spread of disease and to facilitate the growth of their bodies. The pig CAFOs also store the waste material, containing blood, urine, excrement, chemicals, and drugs, in large open-air lagoons.

CHICKENS

Chickens that are raised to produce meat are known as broilers. Broilers are different breeds from the hens that are kept for producing eggs in industrial establishments; a few variations in each class are available. The characteristics that qualify a breed to be a broiler are heavy breasts and thighs, since these are the most desired cuts of meat. Genetic selection and regulation of diet have reduced the time from birth to slaughter in broilers to about six weeks, a small fraction of the time it took to ready the birds for market a few decades ago.

Soon after hatching, broilers are transferred to a large building, typically 40 feet wide and 500 feet long. About 20 to 30 thousand birds are crowded together in these buildings, which have feeding dispensers along their length. Most large facilities have a few such enclosures, each housing birds at different stages of growth for a continuous supply of meat, so that the total number of birds raised at any given time may be as high as 200,000 to 300,000. The younger chickens have some room to move around, but the space per chicken decreases as the birds grow; eventually, there is less than one square foot of space for each bird—not even enough room for it to spread its wings. The broilers only leave these warehouses when it is time for slaughter.

Chickens are fed a mixture of corn and soybeans as well as the ground-up animal by-products that are not given to cattle because of the concern about mad cow disease. Waste material from the fishing industry is often included in the feed of chickens. Chicken feed sometimes includes Roxarsone, an antimicrobial drug. Although the poultry industry claims that Roxarsone fights parasites and increases the birds' growth, it also makes the meat appear more pink and attractive. The drug has generated controversy because it contains arsenic, which may cause cancer, dementia, and neurological problems in humans. The U.S. Food and Drug Administration (FDA) allows a level of arsenic of 2,000 parts per billion in food items, whereas the EPA only allows 10 parts per billion in drinking water. A 2004 *Consumer Reports* study found no detectable arsenic in the muscles of chickens, but it found arsenic in some of the chicken livers in levels that, according to EPA standards, could cause neurological problems in a child who ate 2 ounces of cooked liver each week or an adult who ate 5.5 ounces weekly.[9] The arsenic additive also creates health risks for farmers who work with the chemical. The European Union (EU) concluded that any amount of arsenic is unacceptable and outlawed its use in 1999. In 2006, 70 percent of the 9 billion broiler chickens produced annually in the United States were given Roxarsone.[10]

Most chickens raised in factory farms get infected with *Salmonella* and *Campylobacter*. These pathogens become embedded in the large buildings that are used over and over again, and they infect each new batch of chickens. Since they do not cause particular harm to the birds, and the bodies of chickens provide a hospitable environment for them, the pathogens usually survive in the chicken meat that consumers pick up in supermarkets. As a result, more than 80 percent of chickens in supermarkets carry one or both of these organisms.[11] Birds that live outdoors in natural surroundings, the so-called free-range birds, are generally not infected by these organisms.

Chickens raised primarily for the production of eggs are referred to as laying hens, or layers. From the industry's perspective, the most important characteristic of a breed used for laying eggs is that it should produce a large numbers of eggs per year. In 1940, an average laying hen produced 134 eggs per year. Manipulation of genetics, environment, and feed have doubled this output, so that a modern layer produces 250 to 300 eggs annually. Life for these chickens begins in an incubator on a pullet farm. The male chicks of these breeds are considered useless, because they do not develop enough muscle in a short period of time to be economically useful, and they are discarded.

When female chicks reach egg-laying age—about 16 to 18 weeks—they are moved to a laying facility. These areas typically consist of several buildings, each of which may be the length of a football field. Such enclosures have row upon row of wire cages, known as battery cages, that are often stacked as high as eight tiers. The wire floor of each cage is sloped so that the eggs roll down onto a conveyer belt for mechanized collection and packaging. Four to five birds are crammed into each battery cage so tightly that they cannot spread their wings. Such intensive egg production is highly unnatural and causes extreme stress on hens' bodies. In battery cages, hens often peck each other, causing serious injuries. To limit the damage from this behavior, the front part of the beaks of laying hens are cut off (this is done to broiler chickens, as well). After about one year, the productivity of hens begins to decline and they are either slaughtered or put through the painful process of force-molting, which substantially regenerates the cycle of laying eggs. The flesh of laying hens is generally low quality and is turned into chicken stock or low-grade meats.

* * *

The CAFO style of production has transformed the availability and quality of animal-based foods as well as the environmental impact of raising animals for food. With the help of selective breeding, confinement, and formulation of specific diets, the industry is able to increase the output of meat and dairy products by a substantial amount. The major achievement of these facilities is that the animals reach market weight in a small fraction of the time required in natural settings. The 34 million beef cattle slaughtered in the United States each year typically live for 18 to 24 months, while cattle raised in farms may take 36 to 48 months to reach the market and still weigh less. The difference is even greater for the broilers—6 weeks versus 100 weeks! By shortening the lives of farm animals, the CAFO mode of production has substantially reduced the number of animals alive at any given time to produce the same amount of meat. (However, because of increased demand, this efficient production has not resulted in fewer animals overall, just fewer animals at any given time for producing the same amount of meat.) A number of developments make these animal factories possible. The transportation industry plays a major role in these operations. Feeds consisting of corn, soybeans, and other products are grown on farmlands that may be far removed from these facilities. Cattle graze in pastures far from their place of birth in the early phase of their lives and are transported over long distances more than once during their lifetimes. Modern slaughterhouses are designed to handle a very large number of animals in a continuous mode of operation, so animals must travel far to their final destinations. Although the actual area occupied by the CAFOs is small, they depend on vast tracts of land for their operation.

Factory farming began in America in the 1950s, but its pace has been accelerating in recent decades, with both horizontal and vertical integration. Horizontal integration means merging companies that perform similar tasks and that may compete with each other. Vertical integration is achieved when a company's subsidiaries control various phases of the operation, from producing fertilizers for the agricultural farms to bringing meat to the market. Such consolidations are also increasing the sizes of CAFOs. For instance, during the five years between 1987 and 1992, the average number of animals per farming operation increased by 56 percent for beef cattle, 93 percent for dairy cows, 134 percent for hogs, 176 percent for laying hens, 148 percent for broilers, and 129 percent for turkeys. As a result, a handful of companies now dominate livestock production in the United States.[12]

Factory farms have increased the production of meat and other foods of animal origin by housing a large number of animals in a small area and importing feed from distant areas. Their operations have magnified the impact of raising livestock and growing feed on the local and distant environment, and they have created problems that did not exist on traditional farms. These problems deserve careful consideration so that we can fully appreciate how they are likely to determine whether we can sustain our present lifestyle during the coming decades.

2 * ENVIRONMENT

The simple fact is that farm animals adversely affect the ecosystem that supports all forms of life. They consume vegetation, compact the ground, and produce waste material. But whether their impact is sustainable or catastrophic depends on a number of factors. While the environment can support a small number of animals without causing long-term damage, a high density of livestock may obliterate features of the landscape for a very long time. Confining tens of thousands of animals in a small space in the Concentrated Animal Feeding Operations described in chapter 1 damages the environment by orders of magnitude greater than raising a few animals on a family farm and creates other problems that do not exist in small-scale operations. The direct impact of CAFOs starts in the immediate vicinity of these facilities and spreads outward; producing and obtaining the resources needed to run these operations have an effect that sometimes extends even to distant areas.

ANIMAL WASTE AND THE ENVIRONMENT

In the traditional, mixed farming system, animal waste is an asset! The waste is plowed back into the land so it can decompose. In the process of decomposing it releases essential minerals and nutrients so the next generation of plants and other organisms can benefit from them. Helpful bacteria, fungi, protozoa, arthropods, and earthworms living in the soil degrade the waste produced by humans, livestock, and wild species. This natural cycle cannot keep up with the enormous amount of waste produced by large operations, however, and therefore the waste becomes not an asset but a major liability and a source of multiple problems.

On average, farm animals produce waste at the rate of 15 to 30 times their weight annually; thus for each 1,000 pounds of the animal's weight the animal will produce between 15,000 and 30,000 pounds, or 7 to 15 tons, of waste each year, depending on what kind of animal it is. A hog produces 2 to 4 times as much waste (in weight) as a human does, while a dairy cow produces waste weighing as much as the waste of 23 people. CAFOs generally mix the animals' solid and liquid waste together and then flush it away from the animals using water and mechanical systems or electrical pumps. A farm with 10,000 breeding hogs produces more than 6 million gallons of liquid waste per year. The waste produced by the same number of dairy cows and beef cattle on feedlots will be greater by factors of 9 and 4, respectively. According to the U.S. Department of Agriculture, livestock produce 1.6 billion tons of waste in the United States annually.[1]

This animal waste is stored in the vast open-air pits called lagoons described in chapter 1. Some of these lagoons cover six or seven acres and hold 20 to 45 million gallons of waste products. They are a common feature of the growing number of factory-sized animal operations located in more than 30 states across the

Table 2.1. Manure produced by livestock

Livestock	*Number of animals per 1,000 lbs (Animal Unit)*	*Tons of manure per Animal Unit*
Beef cattle	1	11.5
Dairy cows	0.74	15.24
Breeding sows	2.67	6.11
Other hogs and pigs	9.09	14.69
Turkeys	50	9.12
Laying hens	250	11.45
Broilers	455	14.97

Source: U.S. Department of Agriculture, Natural Resources Conservation Service, www.nrcs.usda.gov/technical/ECS/nutrient/animalmanure.html.

United States and in many countries around the world. In addition to the manure, the lagoons also contain afterbirth, blood, and other animal products. These biological mixtures are conducive to the growth of pathogens, including bacteria, viruses, and parasites, which, in addition to causing risk to human health, emit noxious and poisonous gases in the process of breaking down waste.

The National Institute of Occupational Safety and Health (NIOSH) of the U.S. Centers for Disease Control and Prevention (CDC) has determined that the gases accumulated around these lagoons deplete oxygen from the atmosphere, include explosive gases, and are toxic to higher forms of life. The lagoons are stinking public health hazards. For any worker who accidentally falls into them, or is exposed to their emissions for extended periods, the lagoons are potentially deadly. Records maintained by the U.S. National Institutes of Health show that 19 people died due to hydrogen sulfide emission from manure pits in 1998.[2] Although there are thousands of such facilities in major agricultural states in the United States, the general public is not aware that they exist or that they pose health and environmental hazards, because the facilities are located in sparsely populated areas and access to them is restricted mainly to the workers.

Most developed countries have rules and regulations to control the effluents from large animal operations, but rules are often circumvented or violated. The U.S. Clean Water Act of 1972 specifies that CAFOs are industrial sources of pollution that must obtain permits before discharging into lakes, rivers, or streams. The U.S. Environmental Protection Agency (EPA) has largely failed to implement and enforce these requirements, however, and only about 19 percent of the 13,000 operations in the country have followed the procedure to obtain a permit. Spills and leaks from lagoons are very common, even expected. The state of Iowa authorizes a legal leakage rate of 15 million gallons annually from a lagoon that occupies seven acres of land, but many lagoons leak waste products at rates above the legal limit.[3] From 1995 to 1998, at least 1,000 spills or other pollution incidents occurred at livestock feedlots in a total of 10 U.S. states.

Many lagoons leak because they are not lined, but leaks occur even when liners are used, and seepage rates may be as high as millions of gallons each year. When a CAFO abandons a site because it has become excessively polluted or for other reasons, the abandoned lagoons pose a threat to the environment and water quality for a long time. Thousands of abandoned lagoons exist all over the country. There were 600 abandoned pits in the state of North Carolina alone in 1999.[4]

In some instances the incidents go well beyond leakage or seepage. Lagoons filled with manure have spilled and burst, dumping thousands and often millions of gallons of waste into rivers, lakes, streams, and estuaries. In 1995, after heavy rains accompanying a hurricane, two lagoons burst in North Carolina, releasing 34 million gallons of animal waste into nearby bodies of water. The EPA documented 329 spills in Iowa between 1992 and 2002. Numerous reports of leaks of animal waste from various facilities in the country are documented each year, some of them resulting from deliberate actions, presumably to get rid of the waste.[5] The number of reported incidents of spills and leakages has been steadily increasing in European countries as well.[6] Every year, millions of gallons of waste spill from lagoons into rivers and wetlands, killing fish and other aquatic life and contaminating drinking water.

Applying Waste to Farmlands

Disposing of the enormous amount of waste produced by these animal factories is a major challenge for the CAFOs. Transporting the voluminous semisolid manure is expensive, so untreated waste is commonly applied to croplands and pastures located close to the CAFOs. Large sprinklers transfer the waste from trucks to land using a method known as the sprayfield system.

But as it becomes more difficult to find farmland that can profitably assimilate the nutrients in the sludge, and as the volume of waste continues to grow, CAFOs run out of options—and many of them resort to improper practices. For example, more waste may be applied to farmland than the crops require. Or the liquefied waste may even be applied to nitrogen-fixing crops such as soybeans and alfalfa that require little or no additional fertilizer.[7] Operators have sprayed waste in windy and wet weather, on frozen ground, and on land already saturated with manure. Excess manure can inhibit the growth of crops, contaminate soil, cause surface and groundwater pollution, and waste valuable nutrients.

Polluting Groundwater

It has been estimated that only about half of the nutrients in livestock waste applied to soil are incorporated into crops;[8] the remaining nutrients and other components end up polluting the air, water, and soil, both in the vicinity of CAFOs due to leaks and spills and on the agricultural farms where it is indiscriminately applied. When lagoons overflow or leak, wastewater seeps into groundwater, and the pollutants eventually flow into local bodies of water. Even when the lagoon itself does not break, the groundwater may become contaminated because of cracks in the lining or in the pipes and hoses connected to the lagoons. In all such events, the groundwater becomes overloaded with both nutrients, such as nitrogen and phosphorus, and also pathogenic microbes that may have started proliferating in the energy-rich medium of the lagoons.

Around the world, approximately 8.3 million tons of nitrogen in the form of nitrates and 2.5 million tons of phosphorus in the form of phosphates from manure end up contaminating freshwater resources.[9] Although these nutrients, in the right combination, are essential for the life of plants and animals, excessive amounts harm all forms of life and create serious public health risks. Nitrogen levels above 10 mg per liter in drinking water increase the risk of methemoglobinemia, also called blue baby syndrome, a form of infant poisoning in which the blood's ability to transport oxygen is greatly reduced, sometimes to the point of death. High levels of nitrogen have also been linked to spontaneous abortions and cancer.

Oxygen dissolved in water is essential for aquatic life; most species of aquatic organisms will die in the oxygen-deprived hypoxic zones. High concentrations of organic substances and nutrients caused by runoff from lagoons reduce the amount of oxygen in water, because decomposing organic matter consumes oxygen. According to the EPA's 1998 national Water Quality Inventory, 30 percent of rivers, 44 percent of lakes, and 23 percent of estuaries surveyed suffer from nutrient pollution.[10]

Excessive amounts of nitrogen, phosphorus, and other nutrients foster the growth of a type of algae known as *Pfiesteria piscicida,* which is poisonous to fish. This toxic algae grows in vast quantities in many lagoons, which means that runoff of water from these facilities frequently kills millions of fish in neighboring bodies of water. Between 1995 and 1998, 200 manure-related fish kills occurred, resulting in the death of 13 million fish.[11] According to the EPA, runoff from agricultural activities, including CAFOs, are responsible for degrading the nation's waterways to the extent that 40 percent of these waterways do not meet the safety requirements for fishing, swimming, or both.[12]

CAFOs also create a substantial burden on the local supply of groundwater because a large quantity of water is needed to flush the manure into the lagoons and to service the animals. How much liquid will seep from a lagoon or sprayfield into the groundwater depends on the location of the lagoon or sprayfield. Because water is essential for these operations, many of them have been permitted to set up in close proximity to waterways and floodplains. Their proximity to water alone increases the likelihood of ecological damage; chances of groundwater contamination are greater if a CAFO is located on alluvial soil or where the groundwater runs close to the surface.

Releasing Toxic Chemicals and Heavy Metals

CAFOs are a fertile breeding ground for parasites and nonbeneficial insects because of the nutrients and filth that accumulate in them. To combat these harmful intruders, these facilities apply insecticides and pesticides at frequent intervals. Some insecticides are also fed to animals to remove the worms that develop in their bodies because of their unnatural living conditions. Most of these insecticides and pesticides are injurious to human health; the two most frequently used classes of insecticides—organochlorine and organophosphorus—are known to cause cancer in humans. According to data compiled by the USDA, livestock producers in this country applied 2.16 million pounds of these chemicals in cattle operations in 1999, 72 percent in beef CAFOs and 28 percent in dairy CAFOs.[13] Other operations on factory farms involve using dioxin and polychlorinated biphenyls (PCBs), both of which may eventually find their way into food eaten by the public. PCBs are known carcinogens, and even low levels of exposure to dioxin can adversely affect human reproduction and development and thus diminish the health of infants and children.

A common industrial practice is to add chemicals containing metallic compounds to the rations of livestock, either to stimulate the animals' growth or to kill pests. The metals commonly added to the feed include copper, zinc, selenium, cobalt, arsenic, iron, boron, molybdenum, and manganese. In the pig industry, copper is used because it acts as an antibacterial agent in the gut, and zinc is used to control post-weaning diarrhea. Some of the heavy metals and salts fed to animals are eventually transported to neighboring areas via wastewater. Additionally, heavy metals accumulate in the solid sludge in the bottom of lagoons, where they reach toxic levels after about 20 years or so, at which time the sites are usually abandoned.

Most heavy metals in animal manure present risks to human health and to the biosphere. Manure runoff contaminated with trace elements can end up in bodies of water, where the metals become increasingly concentrated as they make their way up the food chain, eventually reaching fish and crustaceans eaten by the public. The accumulation of heavy metals in sediments, aquatic organisms, and plant and animal tissues harms the reproductive and immune systems of many aquatic and bird species. The human illnesses associated with high levels of trace elements include skin and internal organ cancers. Excessive amounts of arsenic lead to vascular problems and liver dysfunction, selenium can cause hair and nail loss, and too much zinc causes iron deficiency anemia.

HARMFUL MICROBES AND PATHOGENS

The change from raising livestock on small mixed farms to raising them on large-scale factory farms and the resulting establishment of huge slaughtering and processing facilities have created ideal breeding grounds for pathogens, and these organisms have many opportunities to enter the food chain. Notable among bacteria that are harmful to human health are several species of *Campylobacter*, *Escherichia coli* O157:H7, several species of *Salmonella*, and *Clostridium botulinum*. *Campylobacter* species are responsible for about 10 percent of all recent cases of diarrhea in the United States, and approximately 73,000 cases of infection from *Escherichia coli* O157:H7 are reported each year. About half the chickens and turkeys sold in the country are infected with *Salmonella*.[14] According to the CDC, for the period 1998 to 2002 there were well over 1,000 outbreaks of foodborne diseases each year, affecting 128,370 persons. Of the outbreaks for which causes could be established, 55 percent were caused by bacterial pathogens, 33 percent by viral pathogens, and 10 percent by chemical agents.[15]

Most people who become infected with pathogens that thrive in CAFOs are infected by eating tainted meat. When animals are slaughtered, their meat becomes contaminated with the bacteria residing in their digestive tracts. These pathogens may survive in meat if it is not properly cooked or contaminate other food items through contact. Pathogens are also transported to neighboring regions through air and water, and they can infect humans this way, as well.[16]

Superbugs

One of the most dangerous recent developments, one that has the potential to affect the health and welfare of a large number of people, is the growth of "superbugs." Superbugs are new strains of infectious bacteria that are resistant to antibiotics that successfully treat the old strains. This means that there is no cure for the infections caused by the new strains.

As noted in chapter 1, antibiotics are regularly included in the feed of livestock in CAFOs to treat illnesses, prevent infections, and fatten animals on less feed. For decades, farmers have mostly had free rein in dosing livestock with antibiotics. The Union of Concerned Scientists has estimated that at least 24.6 million pounds of antimicrobials are given to healthy animals each year for non-therapeutic purposes.[17] In fact, these scientists estimate that 87 percent of commonly used antibiotics are given to farm animals and only 13 percent are used to try to cure illnesses in humans.

Regularly administering antibiotics to animals in amounts smaller than those required to cure diseases eventually leads to the development of resistant strains. Bacteria multiply very rapidly, usually doubling in number in a matter of minutes, and random natural genetic mutations often create strains that are somewhat different from the main bacterial population. In the course of time, such mutations will produce a strain that is resistant to the existing antibiotic, and this strain will then multiply and spread, unaffected by the antibiotic.

A substantial portion of drugs given to livestock do not degrade in their bodies and end up in the waste lagoons, which then act as mixing vessels and growth chambers for antibiotic-resistant strains. The resistant strains that develop are called superbugs because of their seeming invulnerability—they cannot be wiped out with the drugs that normally kill these bacteria. When the contents of the lagoons are spread on crops and other farmlands, the groundwater and runoff carry both the drugs and the resistant bacteria to soils and waterways, and from there they may infect large numbers of people. Antibiotic resistance in pathogens is a public health problem of increasing urgency.

Several studies have confirmed that poultry and ground meat products are routinely contaminated with pathogenic bacteria, a significant percentage of which are resistant to one or more antibiotics. A 2002 study found that 37 percent of the broilers found in major grocery stores were contaminated with antibiotic-resistant pathogens.[18] Spokespersons for the meat industry usually maintain that antibiotic-resistant strains develop in hospital settings and that the administration of antibiotics has a minimal effect, if any, on this dangerous trend that is rapidly exhausting the arsenal of available drugs to fight infectious diseases. Although drug use on animal farms may have little to do with drug-resistant tuberculosis or other pathogens transmitted among people, careful investigation by scientists has shown that the overuse of antibiotics by the livestock industry is responsible for drug resistance in foodborne pathogens such as *Salmonella* and *Campylobacter.*[19] Antibiotic-resistant bacteria have been found at higher concentrations in soil and air from livestock facilities that regularly use antibiotics in feed compared with facilities that do not use such drugs.

Since bacteria of different types are known to exchange genes, another danger is that resistance to drugs may be passed on to other kinds of bacteria, creating a situation in which superbugs developed on animal farms would plant the seeds of resistance into even more dangerous pathogens. To avoid this potential catastrophe, governments worldwide are cracking down on the use of drugs in livestock. The first move came in 1969, after a British panel recommended banning growth-promoting antibiotics that cause resistance to drugs used in human medicine. The panel's advice was partially heeded in Europe, where key antibiotics like penicillin and tetracycline were taken out of agricultural use in the 1970s.[20] The EU has issued rules limiting the use of several antibiotics in livestock. Although the U.S. Food and Drug Administration has proposed similar regulations, no decisive steps have been taken in this direction, primarily due to opposition from owners of large animal factories. Pharmaceutical companies also oppose regulations that will severely limit the sale of their products. A bill to limit the use of antibiotics in animal feed (HR962) was introduced in the 110th U.S. Congress in February 2007 but expired without Congress having taken any action. Similar bills were introduced in 108th and 109th Congresses, but these bills were not even assigned to committees for discussion.

The Dreaded *E. coli* O157:H7

Escherichia coli (*E. coli*) is a bacterium that is commonly found in the gut of humans and warm-blooded animals. There are a number of strains of this common bacterium, most of which are harmless. However, one strain, known as *E. coli* O157:H7 or enterohaemorrhagic *E. coli* (EHEC), can cause serious health problems in humans. EHEC produces toxins that cause abdominal cramps and diarrhea, which may progress to bloody discharge, fever, and vomiting. Ten percent of infected people develop serious complications, and 3 to 5 percent eventually die after being infected with EHEC.

Humans get infected with EHEC primarily by eating contaminated foods such as raw or undercooked ground meat products and raw milk. The pathogen's significance as a public health problem was recognized after an outbreak in the United States in 1982. Numerous instances of contamination of meat by this dangerous pathogen have been recorded; it is estimated that *E. coli* O157:H7 causes 73,000 infections and 61 deaths in the United States each year. From time to time the government orders beef, beef products, and other foods contaminated with this dreaded bug to be recalled from the marketplace, and millions of pounds of contaminated food items have been removed from stores in recent years.

This dangerous strain of *E. coli* grows and thrives in the acidic environment created in cows' stomachs when they are kept on a diet of grains instead of hay and forage. These pathogens remain in the digestive tract of cattle and also survive in the lagoons where the waste is stored. When cattle are slaughtered, bacteria from the stomach contaminate the meat and travel all the way to grocery stores. In 2003, the *Journal of Dairy Science* noted that up to 80 percent of dairy cattle carry O157:H7.[21] Food safety measures and thorough cooking prevent these bacteria from infecting most of the people who consume meat and milk products. When cattle in feedlots are switched to diets based on hay and forage, EHEC disappears in only two or three weeks.

In addition to meat and meat products, foods such as spinach and lettuce—vegetable products that are not normally associated with farm-raised animals—have carried this pathogen to humans. According to the CDC, in October 2006, 199 persons became infected with *E. coli* O157:H7 by eating spinach; 102 of them became ill enough to require hospitalization. Thirty-one of these people suffered kidney failure, and three of them died as a result of the infection. An *E. coli* outbreak in December 2006 from food served in a fast food restaurant sickened at least 22 persons, 2 of them seriously. In many cases, the contamination of spinach or lettuce with this dangerous pathogen originates from the practices of the animal factories. This strain of bacteria survives and grows in the animal waste stored in the lagoons; if the contents of these lagoons are applied to the farms at inopportune times—such as when the produce is about to be harvested—the plants become contaminated.[22] The bacteria can spread more easily through salad greens because they do not undergo the cooking process that would usually kill these pathogens.

Avian Flu

One of the major threats on the public health horizon is avian influenza, or bird flu, a form of viral infection that is highly contagious among birds such as chickens, ducks, and turkeys. Infected fowl become very sick and die within a few days. Although humans are generally immune to bird flu, that is not the case for other animals. Among livestock, pigs are infected by strains of viruses that attack birds as well as those that sicken humans; this means that pigs can act as mixing vessels in which different strains of influenza combine to create mutant strains that may then be passed on to the human population. If the new strain of influenza is highly injurious to human health, and also very contagious, it may cause an epidemic.

One type of avian flu that has crossed the species barrier is called H5N1. The mortality rate for people infected with this type of influenza is a staggering 50 percent. Although the virus is capable of causing a global pandemic, at present it is not highly contagious and has resulted in only a few fatalities in Asia, Eastern Europe, the Near East, and Africa. However, the virus may become more contagious to humans if a human or a pig is infected simultaneously with H5N1 and with a strain of influenza that spreads rapidly among the human population. A more contagious H5N1 could cause a pandemic, an epidemic of global proportions. In this regard, we need to consider that the influenza pandemic of 1918-19 killed 25 million people in the first 25 weeks; estimates of the total number of people killed by that virus in various parts of the world range from 40 million to 100 million. Because international travel is now common, a pandemic could cause devastation all over the globe in a very short time. To avoid this scenario, millions of birds are routinely killed at the first suspicion of avian flu.

In 2009, there was an outbreak of a new strain of influenza virus known as swine flu. It evolved due to the mixing of various strains of the disease that sicken pigs, birds, and humans and is a variant of an ancient strain known as H1N1. More than a million persons were infected by the virus on all inhabited continents, and thus a pandemic was declared by the World Health Organization in June 2009. Although the human fatality rate from this virus was not very high, the possibility that a new strain will develop—one that is more lethal—cannot be ruled out. The industrialization of livestock production, which involves widespread use of antimicrobial drugs, has increased the risk that new and virulent strains of microorganisms will emerge.

HORMONES

Hormones are administered to livestock to control their growth, metabolism, and bodily functions. The hormones given to the animals are the equivalents of natural hormones—estrogen, progesterone, and testosterone, and two other synthetic hormones. In the United States, almost two-thirds of beef cattle are treated with hormones, either by injection or through implants. When hormones are used, cattle weight increases 8 to 25 percent daily on 15 percent less feed.[23] Hormones can boost animal growth by 20 percent and cost only a few dollars per head of cattle. About one-third of dairy cows are given BGH to increase milk production. Hormone PG600 has been approved in the United States for use in swine.

Hormones used to increase milk and meat production survive to a significant extent in the food consumed by humans. A substantial portion of these chemicals also ends up in water and soil, disrupting the endocrine systems of fish and other forms of wildlife and impairing human health. In one study, researchers found that fish exposed to feedlot effluent had suffered significant damage to their reproductive systems. Researchers at Tufts University report that exposure to endocrine-disrupting hormones can increase the risk of breast and ovarian cancer in women and testicular cancer in men and can lower male sperm quality and count.[24] Because of the potential side effects of hormones, including breast and intestinal cancer and premature puberty, the EU banned their use in 1988 and prohibited imports of beef from the United States and Canada because it may contain residues of the hormones given to the animals.

AIR POLLUTION

Research indicates that the malodorous air surrounding feedlots—generated in part by the decomposition process—may contain as many as 170 different chemicals as well as composite particles consisting of dust, water, and substances of animal origin. While some of the gases originate from the region where the animals are housed, many harmful gases and volatile organic compounds (VOCs) are formed when anaerobic bacteria decompose liquid manure held in the lagoons. And when the contents of the lagoons are sprayed on farmlands, these farmlands become sources of harmful gases and particles. The distant and local long-term effects of the emissions from these operations include acid rain, global warming, and health problems in humans.

For example, dissolved ammonia in the liquefied contents of lagoons consisting of fertilizers and animal waste is sequentially broken down by microbes to form various oxides of nitrogen, including nitrous oxide. These compounds have harmful effects on the ozone in both the upper and lower layers of the atmosphere. The optimum balance is high levels of ozone in the upper atmosphere and low concentrations in the lower atmosphere, where ozone is a pollutant. But nitrous oxide decreases the concentration of ozone in the upper atmosphere, and complex reactions between various oxides of nitrogen, urea, and ammonium carbonate produce ozone that accumulates in the lower atmosphere.

Ozone gas plays a dual role in the atmosphere: in the upper stratosphere (which extends from 10 to 31 miles above the surface of the earth), it is beneficial because it blocks the sun's ultraviolet rays from reaching the earth, thus helping to protect us from the sun's harmful effects, including skin cancer. Ozone at lower levels, in the troposphere (up to a height of 7 to 10 miles from earth's surface), is harmful because it leads to the formation of smog. It is also detrimental to the ecosystem because it suppresses photosynthesis in many plant species by absorbing useful components of solar radiation, thus decreasing the yield of crops such as soybeans, cotton, and wheat. Estimates indicate that a 10 percent decrease in ozone in the stratosphere would increase the ultraviolet radiation reaching the earth by 20 percent,[25] and higher levels of ozone in the lower atmosphere may reduce the world's grain production by 10 to 35 percent.[26] Thus the emission of oxides of nitrogen from industrial livestock operations ends up increasing the incidence of skin cancer and decreasing the output of farms.

Nitrogen compounds also make their way into rivers, lakes, and ponds, where they alter the acidity levels of bodies of water. High concentrations of ammonia are harmful to aquatic life-forms and can even kill them. Ammonia also evaporates from water's surface, ultimately returning to earth with rainfall, damaging the ecosystem. Fully 94 percent of the ammonia emitted worldwide each year has its origin in the agricultural sector; livestock operations contribute two-thirds of the total. The amount of ammonia released from manure depends on how it is stored and disposed of. It is released in greater amounts when manure remains on the surface of the land than when it is used as a fertilizer by mixing it with soil.

Ammonia is just one of the substances present in the lagoons that has a deleterious impact on human health and the environment; others include hydrogen sulfide, VOCs, particulate matter, and methane. Methane is toxic at high levels, but people exposed to this gas may not be aware of the danger it poses because it is a colorless, odorless, and tasteless gas. (The effects of methane on climate are discussed below.)

Hydrogen sulfide, emitted by the manure, is a colorless gas that smells like rotten eggs. Toxicity studies have shown that this gas is rapidly absorbed by the lungs of humans; how, and how severely, it affects their health depends on its concentration. Levels in the range of 500 to 1000 ppm (parts per million) may render a person unconscious, cause serious and debilitating damage to their neurological and respiratory systems, and even threaten their life.

VOCs are gases that react with air in the atmosphere to cause air pollution. These gases are produced in the digestive system of cattle and are emitted through flatulence and burps. Humans who are exposed to VOCs may develop headaches, lose coordination, and suffer damage to the liver, kidneys, and central nervous system.

Particulate matter from the lagoons are minute airborne particles composed of dust, water, and animal waste. Animal feeding operations produce copious amounts of such particles when, for example, animals are moved around and high-volume fans are used to circulate air in the buildings. Such particles are also formed when the buildings are cleaned with high-pressure sprays and when the contents of lagoons are sprayed on croplands. The smallest particles cause the greatest amount of damage, because they are deposited in the alveoli of human lungs, decreasing the lungs' capacity to exchange gases with the blood—a life-sustaining function. Larger particles are deposited in the upper airways of the human respiratory tract and also cause long-term damage.

All of the chemicals discussed here persist in the atmosphere, which means that their cumulative effect is slowly increasing with the passage of time.

Greenhouse Effect and Climate Change

The temperature of the earth's surface is regulated by a mechanism called the greenhouse effect. Without it, the temperature would fluctuate wildly between extremes, depending on whether a particular region were exposed to the sun or not, and the average temperature of the planet would be 20°F (–6°C) and not 60°F (15°C), the current average. The earth sends energy received from the sun back to space by reflecting light and emitting heat in the form of infrared rays. A fraction of this heat is absorbed by the greenhouse gases present in the atmosphere, thus providing an envelope that warms the earth and also eliminates large fluctuations in the temperature of the planet (as occur on the moon, which does not have an envelope of air). But there can be too much of a good thing, and too much greenhouse effect leads to global warming. The main greenhouse gases that lead to global warming are carbon dioxide, methane, nitrous oxide, and chlorofluorocarbons.

How effectively greenhouse gases warm the earth depends on

two factors: how well they absorb radiation emanating from the earth and how long the gases persist in the atmosphere. Discussions of the greenhouse effect generally center around carbon dioxide, because it is emitted in amounts that are much greater than other gases. However, methane is 21 times more effective in trapping heat than carbon dioxide, and it remains in the atmosphere for approximately 9 to 15 years. Each greenhouse gas has been assigned a "global warming potential," or "GWP," in relation to carbon dioxide. Methane has a GWP of 23, which means that it is 23 times more effective in warming the earth than carbon dioxide. Nitrous oxide is even more potent as a greenhouse gas, because it is more effective in absorbing radiation and it stays in the atmosphere for about 114 years; its GWP is 296.[27] Even though their concentrations in the atmosphere are much smaller than that of carbon dioxide, these two gases contribute significantly to the greenhouse effect because of their greater potency; methane is causing almost one-tenth of global warming, and nitrous oxide is causing almost one-quarter of global warming.[28]

The average temperature of the earth has risen by 1.6°F during the last century, and the warming rate has been accelerating with time. Scientists on the Intergovernmental Panel on Climate Change now predict that the average temperature of the earth will further increase by 2.5 to 10.4°C by the end of the century. Some effects of global warming are already evident—for example, decreased rainfall in the tropics and stronger hurricanes and more drought in various parts of the world. The water shortage in the southwestern United States is likely to intensify, and temperature and precipitation will be affected almost everywhere.

For reasons that are not fully understood, the rate of warming is much greater at higher altitudes and near the poles—with the result that ice caps and glaciers near the Arctic Circle, Antarctica, and mountain peaks are melting faster than predicted for the measured increase in temperature. Many of these effects are already observable through the melting of glaciers at higher altitudes and higher latitudes. The melting of ice caps on snow-covered peaks is particularly worrisome, because gradual melt from these ice caps is needed to feed major rivers on all continents, and there is concern that the rivers may not flow all year round. In the United States, the effects of global warming are most visible in Alaska, where forests have been lost and "permanent" ice has melted.

Another effect of global warming that will have repercussions around the world is an increase in the level of seas, as warmer water in the oceans expands and the vast store of ice in Greenland and Antarctica melt. The sea level has already risen by 4 to 8 inches, and there is evidence that it is continuing to rise at a rapid rate. Coastal areas of all continents are at risk of losing substantial amounts of land, because they will be submerged. When we consider that nearly a billion people live within three feet of sea level, the potentially devastating consequences become clear.

Changes in climate may well affect the lives and livelihoods of hundreds of millions of people. The projected effects of global warming include changes in the amount of rainfall, with frequent floods and droughts, and also extreme events such as major storms. While rainfall is necessary for farmlands that are not irrigated, excessive rainfall at the wrong times can have a devastating effect on the agricultural productivity of the land. Much of the earth's plant and animal life has evolved during the last few centuries in a delicate balance with the cyclic temperature variation in each region. Because major crops are fine-tuned to the local temperature with limited tolerance for significant variations, a change of climate can have a devastating effect on agricultural output.

Climate change—bringing with it rising temperatures, rising sea levels, melting ice caps and glaciers, and changing weather patterns—is the most serious environmental challenge facing the human race.

Contribution of Livestock to the Greenhouse Effect

Livestock production currently contributes about 18 percent of the total global warming effect caused by all human activities. The contribution of livestock to the emission of greenhouse gases exceeds that from any other human enterprise—including the contribution of automobiles. Whether directly or indirectly, livestock are responsible for much of the atmospheric emission of the three major greenhouse gases: carbon dioxide, methane, and nitrous oxide.[29]

The respiration of livestock accounts for only a very small part of the net release of carbon dioxide; much more is released indirectly from fossil fuels used to sustain livestock operations. For example, fossil fuels are needed to produce mineral fertilizers and other chemicals used on agricultural farms that grow animal feed crops, and fossil fuels are used to transport animals, their feed, and other related items. Many forests have been cut down, especially in tropical countries, to support livestock operations. Deforestation adds to the load of carbon dioxide in the atmosphere because the forests, which act as sinks of this gas, become depleted (deforestation is discussed more fully in the next chapter). All the activities related to livestock operations cause more than 120 million tons of carbon dioxide to be released into the atmosphere every year.[30]

Livestock operations contribute to the greenhouse effect on a huge scale through the release of methane into the atmosphere. Ruminants such as cattle, buffaloes, sheep, goats, and camels emit methane as part of their digestive process when fibrous feed is broken down in their rumens by microbes during fermentation. Swine and poultry also emit a certain amount of methane during digestion. Additional quantities of methane are produced when the manure of these animals is broken down by microorganisms under anaerobic conditions, mostly when manure is kept in a liquid form in lagoons or holding tanks. Manure directly deposited in fields and pastures, or otherwise handled in a dry form, does not produce a significant amount of methane. The amount of methane produced also depends on the manure's energy content, which is greater when animals are fed grains in feedlots than when they are raised on hay and forage. The total amount of methane released from livestock-related activities has been estimated at 55 million tons per year.[31]

The third major contributor to the greenhouse effect that is produced by livestock operations is nitrous oxide, N_2O. Almost *two-thirds* of the nitrous oxide in the atmosphere has its origin in livestock operations. This potent greenhouse gas is emitted through the unmanaged waste of grazing animals, the manure kept in lagoons, and the application of manure to farmlands. The contribution of nitrous oxide to global warming is more than twice that of methane.

* * *

The large number of animals that are being raised to meet the demand for foods of animal origin puts enormous strain on the planet's resources and its environment. This stress is increased by the practice of keeping livestock in factory-type operations. There are so many problems created by this industry that it is impossible to address them on a piecemeal basis. Instead, a rethinking of the whole system is required.

Food for thought is offered by Gidon Eshel and Pamela Martin of the University of Chicago, who have shown that switching from the typical American diet to one based on vegetable products more effectively reduces the contribution of human food consumption to the greenhouse effect than replacing a typical family car with a hybrid car.[32] Although the livestock industry's effects on the environment and human health represent only a part of the total adverse effect of this industry, those effects alone justify taking corrective actions to reverse the trend of consuming meat as a major part of our diet.

3 * LAND

All forms of life on the earth depend on the resources of the ecosystem for their sustenance. The energy of the sun, converted by plants into useful products with the help of air, water, and nutrients, is cycled through various interdependent organisms. A dynamic balance is maintained among species, with frequent adjustments being made to accommodate changing circumstances. Human actions may upset the natural balance by making excessive demands on some components to satisfy their immediate needs. The resulting disturbance of the ecosystem may be serious enough to endanger its fecundity and may result in permanent or semipermanent damage.

The number of farm animals raised these days is driven by the demand for meat, milk, and eggs. The livestock industry has become so large that it is a major factor in environmental degradation, and the increasing demand for its products has resulted in continuous growth of the industry. As described in chapter 2, livestock consume huge amounts of agricultural items for their growth and sustenance and appropriate a large proportion of the planet's resources. The land must meet their needs in two ways: farm animals graze on vegetation for all or part of their lives, and animals kept in feedlots eat copious amounts of agricultural products.

Degradation of land implies a change in the physical, chemical, and biological properties of the soil—loss of topsoil, excessive salinity, soil compaction, and removal of vegetation—that results in an overall loss of productivity of the land. In lands that are used for grazing animals, degradation occurs because of overstocking of animals, using land at inappropriate times, or using land without replenishing its nutrients. Farmlands that grow grains for livestock suffer permanent or semipermanent damage if intensive farming is carried out, using large quantities of chemicals to grow crops such as grains repeatedly without respite for the land. The resulting loss of fecundity and the accumulation of undesirable chemicals may irreversibly damage the farmland. And the ever-increasing need for grazing lands may result in the razing of forests, which may have ramifications for the climate all over the globe.

In livestock operations everywhere, most cattle, sheep, and goats spend at least part of their lives grazing on traditional farms, rangelands, or pastures. Traditional farms graze their animals and grow mixed vegetation to feed them. Rangelands are usually unimproved lands in arid or semiarid regions that have limited vegetation and cannot support regular cultivation. Farm animals that graze in arid areas are sometimes given water or extra feed by ranchers. Pastures usually have enough water and soil to support low-growing vegetation that is maintained specifically as food for the animals; although these lands may not be rich enough to grow high-value (high-value generally means high-economic-value) crops, they can support agriculture on a limited basis. Overstocking livestock in any of these regions has an adverse effect on their ecology, sometimes with long-term consequences.

DEGRADATION OF RANGELAND AND PASTURES

Most rangelands and pastures can support livestock in a sustainable manner if the stocking density of farm animals is not very high; however, the lands may become severely degraded if used for grazing by so many cattle, sheep, or goats that the land is unable to regenerate its vegetation. According to the International Fund for Agricultural Development (IFAD), the number of cattle on rangelands and pastures throughout the world exceeds the capacity of

Table 3.1. Global distribution of grazing land, in million sq km

Region	*Total land area*	*Grazing land area*	%
North America	18,388	2,730	15
United States	9,167	2,415	26
Canada	9,221	315	3
Central America	2,622	934	35
South America	26,277	7,889	29
Europe	4,727	839	18
Africa	28,545	7,300	26
Oceania	8,286	4,512	54

Source: World Resources Institute, http://earthtrends.wri.org/.

the lands by 70 to 100 percent.[1] Cattle denude the land of vegetation that holds soil in place. When cattle eat all of the useful plants, they are eventually replaced by inedible weeds, thorny shrubs, and unproductive woodlands. Sheep and goats are more destructive in overgrazing than cattle, because they are agile, have a long reach, and will eat leaves and shoots that cattle avoid; these animals can denude land of all traces of vegetation in a short time.

Removing vegetation exposes the ground to more intense solar radiation, which increases evaporation and the risk that plants not eaten by cattle will die from lack of water. The fertility of soil is easily decreased when the vegetative cover is removed. When the land is compacted by the hooves of cattle, rainwater cannot easily soak into the soil. Severe soil compaction, erosion, and decreased soil fertility now affect many cattle-ranching areas, including those in the American West, Central and South America, Asia, Australia, and sub-Saharan Africa.

The amount of rainfall is a major determinant of vegetation and use of land by the livestock industry. Dry lands with limited vegetation cover 40 percent of the earth's land area, exist on all continents, and are primarily used as rangelands for cattle. Among these dry lands, extremely arid lands and deserts have at least 12 consecutive months without rainfall; arid lands receive less than 12 inches of rainfall per year. Semiarid lands may have rainfall up to 24 inches, but it is erratic, with large variations from year to year. Arid lands cover 7.6 million square miles (19.7 sq km) and are used primarily as

Table 3.2. Drylands across the world

Region	*% dryland**
North America	28
Central America	58
South America	32
Europe	24
Asia	39
Africa	43
Oceania	89

Source: World Resources Institute, http://earthtrends.wri.org/.

* Arid and semiarid areas where precipitation is scarce and typically unpredictable.

rangelands; this space is larger than the area of the South American continent and almost twice the area of Europe. Another 5.5 million square miles (14.2 million sq km) of land have sufficient rainfall to support some vegetation and are used as pastures. These are well-managed regions, often with borders and fences, and grow vegetation specifically for consumption by the animals.

Some of these lands could be used to grow crops for human consumption. Land obtained by clear-cutting rainforests, for example, has sufficient nutrients for growing crops but is often used as pasture because of the greater financial returns from livestock. Other pasture land may not be suitable for intensive farming of high-value crops but may still have enough water and sufficiently good soil for growing less demanding crops or trees that produce edible crops for direct human consumption. In the United States, rangelands and pastures represent the largest and most diverse land resources in the country: the grasslands of California, the tundra rangelands of Alaska, the hot and arid deserts of the Southwest, the temperate deserts of the Pacific Northwest, the semiarid cold deserts of the Great Basin, the prairies of the Great Plains, and the humid native grasslands of the South and East. These lands are the primary forage base for the livestock grazing industry in the United States and are utilized by more than 60 million cattle and millions of sheep and goats.

Overgrazing in Arid Regions and Desertification

A very large proportion of land on all continents—roughly one-quarter of the ice-free area of the earth—consists of arid or semi-arid land and deserts. Sixty-six percent of the African continent is arid or desert; 38 percent of the land in Asia is arid or semiarid; Latin America and the Caribbean are about 25 percent dry lands; almost 90 percent of the land in Australia is classified as arid; and 30 percent of the land area in the United States, mostly in the western part of the country, is either arid or extremely arid.[2]

Dry lands are fragile ecosystems that are extremely vulnerable to degradation: their freshwater resources are scarce, and they have shallow topsoil and low biomass productivity. Inappropriately using these lands removes the vegetative cover and degrades the soil, decreasing productivity. In earlier times, when semiarid regions were mostly inhabited by nomads who moved with their herds of animals, their constant movement prevented the land from being depleted.

The earth's surface area, 510 million sq km (*top*).
Distribution of the ice-free area, 131.7 million sq km (*bottom*).

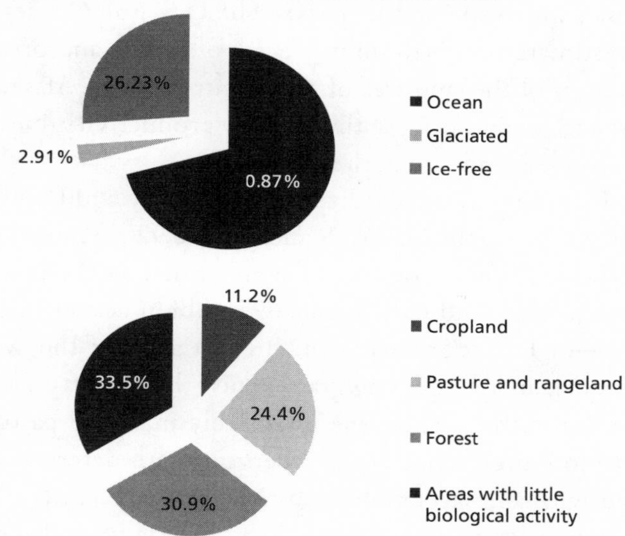

Source: U.S. Geological Survey, www.usgs.gov.

These days, boreholes and windmills make it possible for livestock to stay year-round in areas formerly grazed only during periods when seasonal basins held water. When not correctly planned and managed, provision of drinking water for animals in these re-

gions leads to eradication of vegetative cover from the land, which has no chance of recovering in the rainy season. The U.N. Food and Agriculture Organization (FAO) estimates that overgrazing has caused degradation of 20 percent of the world's pasture and rangeland, but that proportion may be more than 70 percent in arid or semiarid regions.[3] On a worldwide basis, 35,000 to 40,000 square miles (90,000 to 100,000 sq km) are abandoned each year due to extensive degradation of rangelands.[4]

In the United States, where Concentrated Animal Feeding Operations (CAFOs) provide a large proportion of the country's meat, cattle spend the early part of their lives grazing on rangelands. More than two-thirds of the entire land area of Montana, Wyoming, Colorado, New Mexico, Arizona, Nevada, Utah, and Idaho is used as rangeland.[5] Livestock grazing is the most widespread land management practice in western North America. Seventy percent of the western United States is grazed, including wilderness areas, national forests, and even national parks. The U.S. Soil Conservation Service estimates that 410 million acres of public and private lands—21 percent of the land area of the country outside Alaska—are in unsatisfactory condition, with declining productivity. Nearly all of these degraded lands are in the West.

Each year U.S. taxpayers subsidize approximately $100 million to support grazing on public lands. Ranchers use 92 percent—163 million acres (66 million hectares)—of land owned by the Bureau of Land Management, and 69 percent—97 million acres (39 million hectares)—of Forest Service land.[6] In vast areas of the West, domestic livestock are overgrazing grasses and other plants on public lands. Similar to rangelands, the intensively managed pastures in somewhat more productive areas undergo serious deterioration when stocked with animals far above their carrying capacity. The World Resource Institute estimates that half of the land used as pasture in the United States is now overgrazed and suffering from high rates of erosion.[7] A large body of literature is in near total agreement that the world's grazing lands are the most degraded lands on earth.

Continuous degradation of land in semiarid regions may lead to desertification, a process that is characterized by erosion, loss of groundwater, and disappearance of native vegetation. There is a fine line between arid land and desert, and once that line is crossed, it is hard to bring the land back to the former state. Although activities such as cutting wood for fuel may contribute to desertification, it is believed that overgrazing is the major cause of the expansion of deserts in many parts of the world. Desertification converts productive arid lands to wastelands. This is a slow, ongoing process that now threatens 10 million square miles (26 million sq km) throughout the world—just over 60 percent of the rangeland area.[8] According to the United Nations Convention to Combat Desertification, deserts swallow 9,300 square miles (24,000 sq km) of farming land each year.[9] Although countries in the Middle East and Asia are increasingly in danger, EU countries are also under threat. Spain is the most arid country in that region; 66 percent of its surface is vulnerable to desertification. There is a significant chance that Portugal, Italy, and Greece will be mostly arid before 2020.

While 20 percent of the western United States could be termed true, natural desert, perhaps another 20 percent has been so thoroughly and incessantly grazed by livestock that it has taken on the appearance of desert. In its Global 2000 report, the Council on Environmental Quality noted that improvident grazing has been the most potent desertification force in terms of total acreage (352,000 square miles, or 910,000 sq km) within the United States.[10] China has always been arid; about 25 percent of its land mass is composed of deserts. The situation is getting worse, however: old deserts are advancing, and new ones are forming. China is being affected by desertification more than any other major country. According to the Xinhua news agency, 90 percent of China's grasslands have been degraded by overgrazing, a figure that increases by 7,700 square miles (20,000 sq km) each year.[11] China's Environmental Protection Agency reports that the Gobi desert expanded by 20,000 square

Table 3.3. Conversion table, English and metric measurements of land area

1 square mile	=	2.59 square kilometers
1 square mile	=	640 acres
1 acre	=	0.4047 hectares
1 square kilometer	=	100 hectares

miles (52,400 sq km) from 1994 to 1999. The advancing Gobi is now within 150 miles of Beijing.[12] Due to increasing desertification, violent sandstorms that originate in these regions have been battering Chinese cities, and their mustard-colored dust has begun to reach South Korea, Japan, and the west coast of North America.

In northern Africa, increasing exploitation of the regional resources by humans and livestock are extending the area covered by the Sahara desert in all directions. The advancement of the Sahara by 100 kilometers in the Sahel region is primarily attributed to overgrazing. The Great Thar desert in Northwest India does not have a truly arid climate but has been turned into a barren wasteland, mostly due to overgrazing. Similar situations prevail in Kazakhstan, Uzbekistan, and other arid regions of the world. When overgrazed lands turn into deserts, the climate of neighboring regions is also affected. As an essential part of photosynthesis, water in the leaves of plants evaporates—a process known as transpiration. Many plants transpire as much as 90 percent of the water that they absorb with their roots. Elimination of vegetation decreases the amount of moisture in downwind regions. It is estimated that desertification accounts for $42 billion worth of food productivity lost worldwide annually.[13] In addition, there is an unaccountable cost in human suffering due to the hunger and malnutrition that result. The world's natural environment is deteriorating in numerous ways; the degradation of grazing lands is the most pervasive among them.

When arid or semiarid lands are overstocked with farm animals, there is a conflict with wildlife that either compete for resources or threaten the livestock. To facilitate the use of these lands by ranchers, in 1931 the USDA established an Animal Damage Control Program, which was renamed the Wildlife Services Program in 1997. Its main function is to eradicate, suppress, and control wildlife considered detrimental to the western livestock industry. Its agents kill any creatures that might compete with or threaten livestock, such as badgers, black bears, bobcats, coyotes, foxes, mountain lions, opossum, raccoons, and prairie dogs. Intensive ranching threatens the very existence of wildlife in these regions.

Effects of Overgrazing on Riparian Areas in Arid Lands

Riparian areas are narrow strips of land that border creeks, rivers, or other bodies of water. Because of their proximity to water, the plant species and topography of riparian zones differ considerably from those of adjacent uplands. Although the area classified as riparian in these regions may be small, these lands represent an extremely important component of the overall landscape. These zones are particularly important in arid and semiarid regions, because their biological richness sustains all forms of life, particularly plants and wildlife. The availability of water and green vegetation also makes them attractive and important to domestic livestock grazing in these regions. When livestock congregate in these areas in large numbers, the ecological cost of supporting them may be very large and have far-reaching effects.

Livestock change or diminish the vegetation around the rivers, trample the banks, degrade water quality, and increase water temperature. Loss of vegetation and compaction of soil from the hooves of cattle prevent rainwater from being absorbed by the ground. Instead, the flow of water during storms causes high peak flows that erode stream banks and deepen channels. If the soil cannot absorb water, the level of the groundwater drops, decreasing the amount of water available during summer months. If such events continue, they have the potential to dry out riparian areas. Another problem is that nutrients in animal wastes that are deposited in streams cause eutrophication, a process in which aquatic vegetation grows quickly and decomposes, consuming oxygen from the streams and killing all forms of life in the waters. Damage caused by cattle adversely affects all plants and animals and also the overall productivity of these regions.

Although riparian areas comprise less than 1 percent of the land in the western United States, they are among the most productive and valuable of all lands. While all rangelands may enter a spiral of decreasing productivity when they are overexploited, riparian areas are particularly susceptible to damage when livestock are concentrated in them at the wrong time, in too great a number, or for too long. Extensive deterioration of western riparian areas began with severe overgrazing in the late nineteenth and early twentieth centuries. Native perennial grasses were virtually eliminated from vast areas and replaced with sagebrush, rabbit brush, and shallow-rooted vegetation, which are less suited for holding soil in place. When cattle invade springs and rivers in arid lands, the very existence of some extremely rare species of fish, such as desert pupfish and Moapa dace, is threatened. Aquatic remnants from the last ice age, desert pupfish and Moapa dace have survived for thousands of years due to their unique ability to withstand harsh environmental conditions, including high temperatures, high salinity, and unpredictable water flows.[14]

A century of grazing in the intermountain West has degraded streams and riparian systems, stripped uplands of native touchgrass, converted herbaceous to woody communities, increased erosion, and endangered native species. These developments have unleashed natural forces that have literally transformed large areas of the landscape. Grazing has damaged 80 percent of western streams and riparian areas in the United States.[15] Arizona had four major rivers: the Santa Cruz, the San Simeon, the San Pedro, and the Little Colorado. The regions around these rivers were once described as lush and green, but now they are dry along most of their lengths except during violent floods that originate in overgrazed rangeland. California has lost 89 percent of its riparian woodlands since 1848, largely due to ranching, farming, dams, and mining. The states that have suffered extensive damages to riparian regions include Arizona, California, Colorado, Idaho, Nevada, New Mexico, Wyoming, Montana, Texas, Oklahoma, and Kansas.

ENVIRONMENTAL EFFECTS OF INTENSIVE FARMING OF LIVESTOCK FEED

The adverse impact of livestock is not restricted to the lands where they graze. Grains, pulses, and other high-value products constitute a major portion of livestock feed in modern establishments, particularly when animals are brought to feedlots for finishing, as is done in the CAFO mode of production. On a worldwide basis, one-third of the crops grown are fed to farm animals. Adding together all types of lands that are used to support livestock throughout the world, including lands used for grazing and those that grow agricultural items for the use of livestock, almost 30 percent of the ice-free terrestrial area, or about 15 million square miles (39 million sq km), is devoted, either directly or indirectly, to the production of animal-based foods.

As mentioned in a previous chapter, grains constitute the major component of the feed of livestock in the CAFO style of production, thus making farming to grow feed crops an integral component of the livestock industry. Hence the requirements and consequences of growing these crops must be included when considering the environmental effects of dietary choices. Modern varieties of grains and other high-value crops are grown by intensive farming in which the same crop, often corn or soybeans, is grown in a continuous cycle, without the land having a respite so it can regenerate and recover lost nutrients.

In Brazil and the United States, the bulk of cattle and swine feed consists of corn, while in Canada and Europe, wheat and barley are the main ingredients. Feed also contains vegetables and fruit, such as potatoes, cabbage, and plantains, as well as peas and beans. The United States is a major producer of corn; large tracts of land in the midwestern Untied States are dedicated to the production of this single crop. In 2007, U.S. farmers used 80 million acres (32 million hectares) of farmland to produce 347 million tons of corn, more than half of which was used for livestock feed. In the same year, U.S. farmers produced 1.33 million tons of oats, which was not enough to meet the domestic needs of farm animals and had to be supplemented with imports from Canada and the EU.[16] Nearly 80 percent of all soybeans grown anywhere in the world were used in animal feed.[17] The proportion of agricultural land devoted to the production of animal feed is increasing, and intensive farming is degrading the quality of farmlands in numerous ways.

Loss of Topsoil and Degradation of Farmlands

In topsoil, the organic matter that is tightly bound to the inert medium makes the soil fertile. Soil becomes more porous and has a greater capacity to hold water when a substantial part of it is composed of decayed organic matter. Such soil provides nutrients and minerals in a form that can be easily absorbed by the roots of growing plants. For these reasons, topsoil is the most useful component of agricultural land. Fertile soil contains about 40 tons of organic matter per acre (100 tons per hectare) in the form of decaying leaves, stems, roots, and other plant products. In addition to organic matter, good soils are inhabited by some of the world's largest and most diverse populations of earthworms, insects, arthropods, and microorganisms. The combined weight of these living beings in productive soil may be as much as 9,000 pounds per acre. These organisms break down dead plants and animal tissues to form humus, the dark and crumbly carbon-based portion of soil that greatly enhances its productivity. Interactions between the inert soil, organic matter, and minute creatures constitute a dynamic system that preserves and maintains the soil's fertility. Topsoil's importance to agriculture is so great that topsoil is called the foundation of human civilization.

Intensive farming causes a loss of topsoil because frequent turning of soil brings the organic matter to the surface and exposes it to the damaging effects of sun, wind, and rain. Although both rangelands and croplands are subject to soil erosion, croplands are more susceptible to erosion because they are tilled repeatedly and are often left bare for several months between plantings. The organic matter slowly becomes detached and is blown away by wind or removed by water from rain or irrigation systems. Around 40 percent of the soil that is washed away ends up in rivers and streams or deposited in places far removed from farmlands.

Loss of topsoil from agricultural lands is a very serious problem that is diminishing the productivity of farmlands in most parts of the world. The effects of these losses are widespread and long lasting. Worldwide, it is estimated that 75 billion tons of topsoil are lost every year, costing approximately $400 billion, or about $60 for every person on the planet per year (measured by the cost of the food the topsoil could have produced and the cost of synthetic fertilizers that are applied to compensate for the loss).[18] Areas permanently covered by vegetation are substantially protected in a way that farmlands are not, because in untilled soil the biomass is more firmly embedded in the soil, and the energy of rain and wind is dissipated by the plants.

Topsoil loss degrades land through loss of nutrients and by decreasing the soil's capacity to hold water. As the fertile topsoil is thinned, the roots of the plants do not penetrate deep into the soil. The loss of nutrients and lack of accessible water stunts the growth of plants, directly influencing the yield of crops grown there. Soils that suffer severe erosion may produce 15 to 30 percent smaller crop yields than soils that have not been eroded.[19] At present, about 80 percent of the world's agricultural land suffers moderate to severe erosion; agricultural productivity of farmlands has decreased between 15 and 30 percent during the last 25 years, and a significant portion of the land has been abandoned by farmers. In the United States, erosion has reduced the ability of farmlands to produce corn by 12 to 21 percent in Kentucky, 21 percent in Michigan, up to 24 percent in Illinois and Indiana, and 25 to 65 percent in Georgia.[20]

An example of extreme devastation caused by loss of topsoil can be seen in Haiti. Only 3 percent of the once lushly forested terrain has tree cover now, and up to one-third of it, some 365,000 acres (900,000 hectares), has lost so much topsoil that it is no longer arable or barely so. Most of the topsoil was lost in the early eighteenth century, when the land was cleared of native vegetation to grow sugarcane for the export market.

Topsoil loss throughout the world, caused primarily by intensive farming, is a continuous process that will persist in decreasing the productivity of the soil and make it increasingly difficult to feed a world population of seven billion people. Livestock are by far the major contributors to soil erosion on agricultural lands and pastures. As we have seen, food grown to sustain livestock is an integral part of current operations.

Synthetic Fertilizers and Excessive Salinity in Soil

It is possible to partially compensate for the loss of nutrients from soil by using synthetic fertilizers, but added chemicals do not replace all of the lost components and cannot rejuvenate the soil's water-holding capacity. Between 1960 and 1995, the global use of fertilizers increased fivefold and is projected to increase another threefold by 2050. However, chemical products create environmental problems, both in the factories where the fertilizers are manufactured from petrochemicals and in the regions around the farmlands. Plant roots absorb only a portion of the synthetic fertilizers that are applied to the soil. It has been estimated that even under favorable conditions, only 30 to 50 percent of the nitrogen and 45 percent of the phosphorus from the inorganic fertilizers applied to farmlands is absorbed by the roots of the growing plants.[21] The fraction of fertilizers that stays in the soil leads to acidification of the soil, while the rest is washed away with irrigation, entering groundwater and local bodies of water. A significant portion of these nutrients is carried further away by rivers to the gulf regions, where it creates dead zones, areas of the sea where nothing survives because of a lack of oxygen in the water.

Applying chemical fertilizers over long periods and continuously irrigating farmlands increases the amount of salts in the top layers of the soil, causing excessive salinity that is injurious to the plants. The water added to farmland during irrigation contains salts; it also brings other salts already in the soil, or excess salts from the fertilizers, to the surface. A large portion of the potassium in the synthetic fertilizers is in the form of potassium chloride, a salt somewhat similar to table salt. As water evaporates, the concentration of salts in soil increases. Excessive salinity in the root regions of plants makes them less vigorous and lowers the yield of all crops. In extreme cases, it may create a salt crust on the surface of the soil, making it impossible to grow anything. High salt levels are becoming a serious problem in many agricultural areas in almost all countries. Excessive salinity is a greater problem on irrigated farms than on rain-fed farms. Since irrigated farms are much more productive, the loss of output is much greater.

Worldwide, some 20 percent of irrigated land, about 175,000 square miles (450,000 sq km) is salt-affected, including some of the most productive regions of the United States, China, and India. The situation is getting worse: it is estimated that 1,000 to 2,000 square miles (2500 to 5000 sq km) of irrigated land are taken out of production each year due to a large concentration of salts in the soil.[22] Dry lands are more easily affected by increased salinity, which means that an estimated 3.5 million square miles (9.5 million sq km) of salt-affected land occur in arid and semiarid regions, nearly 33 percent of the potentially arable land in the world.[23] The continuing loss of croplands cannot be compensated for by additional inputs for long and is bound to decrease the availability of food in the world.

Deforestation

Though most of the land throughout the world that can be used to grow crops has already been exploited, the reduced productivity of eroded agricultural lands, the needs of a growing population, and the increasing demand for animal flesh make it necessary to continually search for additional areas to grow crops or to serve as pastures for animals. There is a problem, however: the unused land is too steep or too dry or the soil layer is too shallow to use for growing grains on a regular basis. The only way to increase the area of croplands is to cut down rainforests, as is being done in the tropical regions of South America and East Asia.

Although they are essential parts of productive ecosystems, forests are not being protected but rather cleared to replace degraded agricultural lands. Forests sustain various forms of life and contribute to human welfare in a number of ways, including preventing erosion of soil, maintaining soil fertility, and absorbing carbon from the atmosphere through photosynthesis. They also host a large proportion of the earth's biodiversity (see below), protect water catchments, and moderate climate change. Tropical forests, covering less than 7 percent of the earth's land area, contain about half of the earth's species. The Brazilian Amazon, for example, provides habitation for numerous mammals, birds, and reptiles and thousands of types of insects and microorganisms, many of them as yet unknown to science.

The demand for meat is putting severe stress on this precious resource. Latin America has suffered the most dramatic loss of forests; since 1970, farmers and ranchers have converted much of the region's tropical forests to pastures, and from 1990 to 2000, 180,000 square miles (47 million hectares) of forest have been destroyed.[24] Large-scale deforestation contributes to climate change in a number of ways. When living plants are cut down and burned, they release the greenhouse gas carbon dioxide. The removal of organic growth decreases the evaporation of water into the atmosphere, thus affecting precipitation in downwind regions. The vegetative growth, soil, animals, insects, and microscopic life in a rainforest combine to form a living, interdependent system. When trees are cut down, nutrients from the soil are washed away with the rains. The land becomes barren and loses its productivity in just a few years.

Deforestation is affecting Brazil more than any other country in the world. From 1990 to 2000, the country lost about 77,000 square miles (200,000 sq km) of rainforests. Almost 80 percent of this land was converted to pasture, while only 20 percent has been used to grow crops. The amount of beef exported by Brazil increased fivefold during that decade. The cattle industry is expanding at a phenomenal rate in that part of the world; the number of cattle more than doubled, from 26 million in 1990 to 57 million in 2002. In 2003, Brazil exported nearly 1.2 million tons of beef to Western countries, the bulk of it going to the United States, suggesting a strong correlation between deforestation and the growth in international demand for Brazilian beef.[25] In Central America, 40 percent of all rainforests have been cleared or burned down in the last 40 years, mostly to create cattle pastures and meat for the export market.[26]

Monocultures and Loss of Biodiversity

The term biodiversity refers both to genetic diversity—that is, variations in genes in a type of plant or organism—and to species diversity, the variety of organisms living in a region. An enormous diversity of animals, plants, insects, marine life, and other organisms have evolved over millions of years to thrive in a wide range of habitats. Biodiversity creates an intricate network that maintains oxygen in air, enriches soil, purifies water, protects against flood and storm damage, and regulates climate. Throughout history, humans have tapped into this biodiversity for food, shelter, medicines, and cultural and aesthetic pursuits. Many of these living things go unnoticed while they interact with each other to maintain the viability of the ecosystem.

A single tree or cubic yard of soil contains thousands of species of organisms whose functions often are not properly understood or appreciated. Although a significant proportion of drugs are derived, directly or indirectly, from biological sources, and about 40 percent of the pharmaceuticals used in the United States are manufactured using natural compounds found in plants, animals, and microorganisms, only a small proportion of the total diversity of plants has been thoroughly investigated to find out whether they might be beneficial as medications or in other ways. In addition, a wide range of industrial materials, including building materials, fibers, dyes, resins, gums, adhesives, rubber, and oil, is derived directly from biological resources. The most important function of biological diversity, however, is to maintain environmental equilibrium and to help nature recover from any catastrophic events by tapping into the storehouse of resources.

Not surprisingly, the biodiversity of our planet is decreasing at an alarming rate. Mostly due to human activities, we are losing species at somewhere between 100 and 1000 times the natural extinction rate. And, again, our reliance on farm animals for food is largely to blame. The billions of farm animals being raised contribute to the loss of biodiversity in a number of ways, such as substantially reducing the vegetation that supports other animals. In addition, animals that would compete with livestock for food and water are poached or eliminated, so the number and variety of wild animals has decreased. These factors, combined with deforestation, have devastated the habitats of a wide range of species, causing a decimation of native flora and fauna.

Monoculture in the agricultural world means devoting large tracts of farmland, often extending over dozens or hundreds of square miles, to the production of a single crop, usually one of the hybrid varieties. Farmers have to resort to monoculture to meet the increasing demand for grains and other crops. Monoculture is an important factor contributing to the loss of biodiversity on the planet. With a premium on agricultural output, less productive varieties are not grown. For corporate factory farmers, efficiency increases when extensive areas are planted with a single variety of a specific plant, because all of them need the same nutrients and amount of irrigation and can be treated with the same type and quantity of pesticides and chemical fertilizers. The same crop growing over a large expanse of land also makes it possible to more easily sow and harvest with mechanical equipment.

Biodiversity in ecosystems provides crops with resilience to cope with invasion by disease, new strains of pests, or changes in climate or other conditions. It also provides the resources for agronomy research to develop new varieties of plants. For example, when the rice grassy stunt virus struck rice fields from Indonesia to India in the 1970s, 6,273 varieties were tested for resistance to the virus. Only one was found to be resistant, a relatively feeble Indian variety with the desired trait. It was hybridized with other varieties and is now widely grown. In 1970, coffee rust attacked plantations in Sri Lanka, Brazil, and Central America; a resistant variety of coffee was found in Ethiopia, coffee's presumed homeland.

Growing a variety of crops instead of monoculture in a region has other advantages. It may provide a partial safeguard against unexpected changes in weather or rainfall. The seeds that produced the Green Revolution were not synthesized in the laboratories but were produced by crossbreeding existing species. Research to find drought-resistant varieties and those that can tolerate some salinity and incorporate other qualities into plants that may be needed in the future will depend on the availability of a variety of types with different characteristics.

The industrial mode of raising livestock is incompatible with genetic variations in animals because different breeds have different requirements and growth patterns. According to the FAO, 1,000 breeds of farm animals—about 15 percent of the world's cattle and poultry varieties—have disappeared over the last century. About 300 of these losses occurred in just the last 15 years, and many more breeds are in danger of extinction.[27] One example of dependence on a single species involves the leghorn chicken in the United States, where almost all white eggs are produced by this breed.

Insects, arthropods, and other creatures perform crucial functions to stabilize the ecosystem in numerous ways. Insects are also a virtually untapped source of food, dyes, and pharmaceutical products. Various microbes can be effectively used for nitrogen fixation and waste recycling. Earthworms, insects, and fungi are no less fascinating, less useful, or less worthy of attention and conservation than large organisms. But pesticides that are applied to crops also adversely affect biodiversity among these creatures. The amount of synthetic pesticides used worldwide on agricultural crops may be as much as 2.2 billion pounds (1 billion kg).[28] These chemicals cannot be designed to kill a single species of insect, so they harm and kill many species.

* * *

We are reducing the biodiversity of the planet at a very rapid rate through several practices:

- growing monocultures on farmlands, exclusively planting modern hybrid varieties at the expense of indigenous varieties
- raising only those species of livestock that provide the highest return with the minimum investment
- large-scale deforestation
- applying pesticides that kill nontarget as well as target species

These practices are eliminating a precious resource that we have inherited from previous ages and that may be crucial for providing us with food and other resources in the near future. Substantial reduction in biodiversity casts a shadow over the developmental path and perhaps over our very existence because, with our increasing numbers, continuously producing large quantities of grains and other produce is not a choice but a necessity for our survival.

On a worldwide basis, about 34 percent of agricultural land has been lightly degraded, 43 percent is moderately degraded, and 9 percent is strongly or extremely degraded.[29] It has been estimated that 7,500 to 20,000 square miles (20,000 to 50,000 sq km) of agricultural land is lost annually due to loss of topsoil, excessive salinity, and depletion of nutrients.[30] The current trend to consume greater and greater quantities of meat and dairy products in all parts of the world poses a serious threat to the continued productivity of the land.

The number of livestock being produced today is not determined by available resources but by the ever-increasing demand for animal products. This disconnect between the capacity of the planet and the number of livestock being raised has created severe environmental problems and cannot continue for much longer. A simple extrapolation of the present trend indicates that, unless drastic steps are taken to change eating habits, the needs of humans and the decreasing capacity of the planet to provide food will collide, not far in the future, but only a few short decades away.

4 * WATER

Water is a precious resource essential for our survival. Throughout history, civilizations have flourished near sources of freshwater; deserts, for the most part, have been wastelands. In earlier times, when the population of human beings was smaller and lifestyles were simpler, people could simply survive—and thrive—by living close to a supply of water, with easy access to food that grew in neighboring regions.

More than 97 percent of the water on our planet is in the oceans, but the high concentration of salt in seawater makes seawater useless for direct consumption by all forms of terrestrial life, which need freshwater to live. A large portion of freshwater is locked up in glaciers, in permanent ice caps at high altitudes, and in regions around the North and South Poles. Rivers depend on glaciers or permanent ice caps for flowing water throughout the year. Without these sources of water, most rivers would become seasonal, with large variations in their flow depending on the amount of rainfall. Through extensive canal systems, our planet's streams, lakes, rivers, and ponds provide water for irrigation in all parts of the world.

The amount of water required for agriculture greatly exceeds the water required for all other human activities, including household consumption as well as industrial and manufacturing facilities. At least 70 percent of global water is used for growing food; in regions of extensive agricultural activity, such as the American Southwest, more than 80 percent of water is used for irrigating farmlands.[1] If identical crops receive the same amounts of nutrients and sunlight and are grown under similar weather conditions, their yields will depend directly on the amount of water provided to them.

When forage was grown in Utah using only the limited amount of rainwater available in that area, the yield was 13 pounds per pound of applied fertilizer. However, the yield of forage per pound of fertilizer in the southeastern United States, where rainfall was abundant, was 35 pounds, a gain of almost a factor of three.[2] Without sufficient water, application of fertilizers would have only minimal effect on agricultural productivity. The new high-productivity varieties of major crops that ushered in the Green Revolution require even more water than older strains, thus increasing the demand for water. There are clear indications that shortages of water may soon limit the agricultural output of the land.

There are substantial differences in the amount of water required to produce various types of foods, both of plant and animal origin. However, animal products generally require much more water—as much as 10 times more than agricultural products.

Although livestock (except for dairy cows) do not directly consume great amounts of water, Concentrated Animal Feeding Operations depend on a continuous supply of water. For one thing, the waste produced by these animals is mixed with water so it can be easily pumped into the lagoons. Also, cattle are slaughtered in abattoirs that use huge amounts of water to wash carcasses at various steps, including eviscerating and boning. It takes a substantial amount of water to convert a large, living animal weighing half a ton into plastic-wrapped packages that a consumer picks up in the supermarket. For these reasons, the availability of water is one of the main considerations in selecting the locations of feedlots and abattoirs. Poultry processing tends to be even more water-intensive per weight than red meat processing, because water is used to scald the birds so their feathers come off more easily.

It takes huge amounts of water to raise and slaughter animals—and much greater amounts to produce these animals' feed. Since growing grains requires more water than growing forage, industrial animal factories require more water than traditional mixed-farming methods of raising livestock. Corn requires more water than some other grains, further increasing the amount of water used to produce each pound of animal-based food in the United States.

Water is indirectly required in other phases of farming operations, including the manufacture of chemicals such as fertilizers, pesticides, and insecticides and the operation of various equipment used by the farming industry. Taken together, these factors greatly increase the amount of water required to produce foods of animal origin, and thus raising animals for human consumption places a greater burden on planetary resources of freshwater.

RIVERS, STREAMS, AND PERMANENT ICE AS SOURCES OF FRESHWATER

As mentioned above, irrigating, instead of depending on rainfall to provide a regular supply of water to crops, greatly increases the output of farms. Worldwide, only 17 percent of cropland is irrigated, but this irrigated land produces 40 percent of the world's food.[3] In the United States, nearly half the value of all crops sold comes from the 16 percent of the cropland that is irrigated.[4] Since ancient times, rivers, streams, and lakes have been used for irrigation by diverting their water through systems of dams and canals. These sources of freshwater have now been fully exploited. In 1950, there were 5,000 major dams higher than 50 feet; such dams now number more than 40,000. Most major rivers of the world are reduced to a trickle by the time they reach a sea, and many of them dry up before reaching a sea.

As populations increase, so does the need for food. But the need for more food is compounded as more and more people choose to eat greater amounts of foods of animal origin. And as the preceding paragraphs indicate, this trend is depleting freshwater everywhere, and the situation is worsening at an alarming rate. More than a billion people in Africa and Asia have no access to reliable clean drinking water; in many places, women walk for miles to procure water for basic necessities. According to an assessment prepared by the United Nations Environment Programme, about 1.1 billion people lacked access to safe drinking water in 2010, while another 2.4 billion lacked access to basic sanitation.[5] The International Water

The earth's distribution of water, distribution of freshwater, and distribution of surface and atmospheric water (*from top to bottom*).

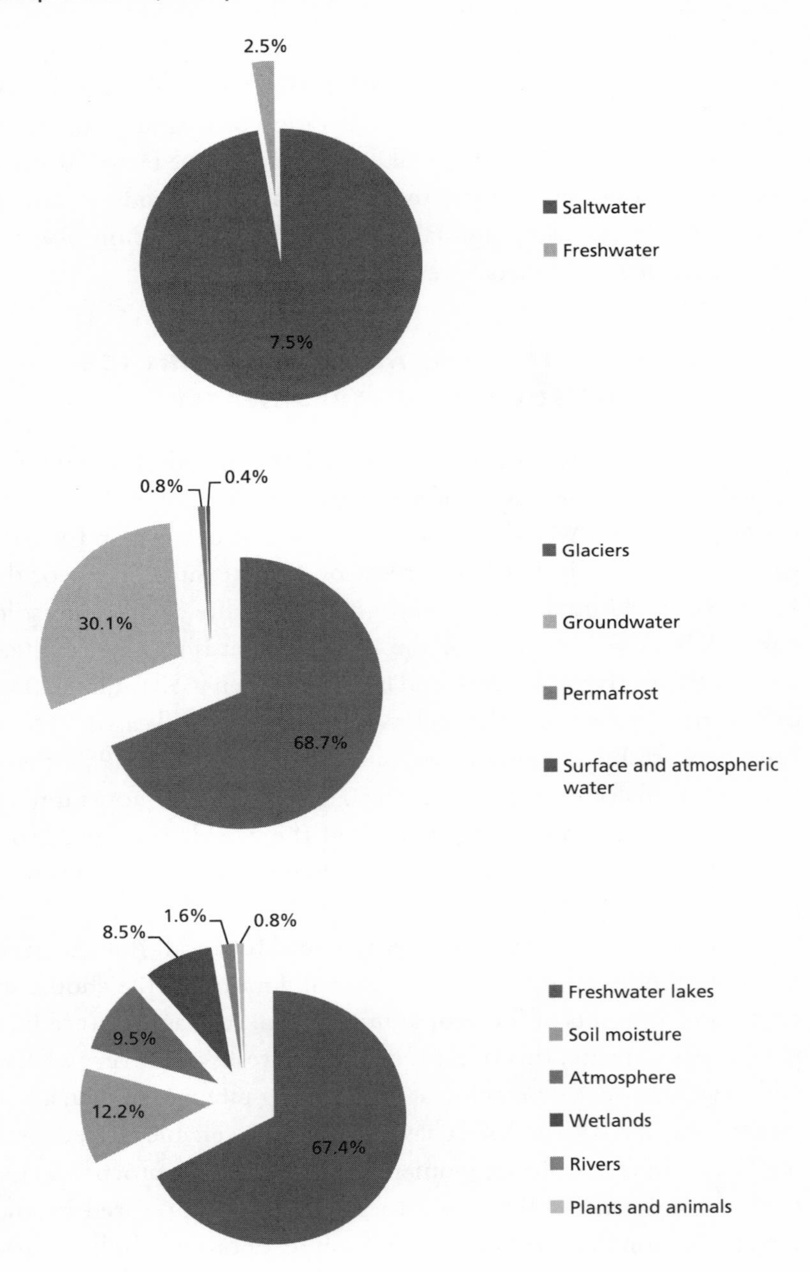

Source: U.S. Geological Survey, http://ga.water.usgs.gov/edu/earthwherewater.html.

Management Research Institute projects that the percentage of the world's population living in water-stressed lands will increase from 38 percent today to 64 percent by the year 2025, with about 1.8 billion people living in regions of absolute water scarcity.[6]

Water levels in some of the world's most important rivers, including the Niger in West Africa, the Ganges in South Asia, the Yellow River in China, the Danube in Europe, and the Nile in Africa, have declined precipitously during the last few decades due to overexploitation, mostly for agricultural purposes. The Nile and Yellow rivers no longer reach the ocean most of the year because so much water is drawn from them upstream. Water levels in most of the lakes are also falling, for the same reason. Several countries, including China, Bangladesh, India, Pakistan, and the Republic of Korea, already suffer from water scarcity, and the shortage of water has severely limited the productivity of their farmlands. Many countries in a band stretching from China, India, Pakistan, and the Middle East to North Africa are severely water stressed and may soon not have enough water to maintain the per capita food production at the present level.[7]

Signs indicate coming shortages of water in just a decade or two almost everywhere, including in the United States. According to the EPA, at least 36 states may face local, regional, or statewide water shortages in just a few years.[8] In Florida, there is an impending water shortage because of the increasing population. In upstate New York, the level of water in reservoirs has been dropping almost every year. The water of the Colorado River, the source of freshwater to seven western states, is used so heavily for industrial, domestic, and agricultural purposes that it does not consistently reach the ocean. The level of water in the Great Lakes, one of the earth's largest aboveground freshwater systems, containing 18 percent of the world's freshwater, is decreasing at alarming rates, as is the water of other lakes in the region.

The Lake Mead and Lake Powell system includes a stretch of the Colorado River in northern Arizona; aqueducts carry water to Las Vegas, Los Angeles, San Diego, and other communities in the Southwest that depend on this essential resource for survival. Tim Barnett and David Pierce of the Scripps Institute of Oceanography have estimated that there is a 10 percent chance that Lake Mead will go dry by 2014 and a 50 percent chance that it will be dry by 2021.[9] The consequences for Southern California are serious.

There are clear indications that the shortage of water will be aggravated by global warming and the accompanying climate change. Crops consume more water with increasing temperatures. Changes in precipitation patterns, some of which are already evident, will lead to a net decline in rainfall. Low rainfall and increasing temperatures will substantially increase the water requirements of the crops, further decreasing the output of farmlands. Climate variability also creates extreme weather patterns that lead to droughts and floods. Whereas droughts create immediate shortages, the water of floods usually flows into rivers and seas and does not compensate for water shortages in times of scarcity.

Glaciers and permanent icecaps are melting almost all over the world at rates that far exceed the scientific predictions. Glaciers are melting in Asia's Himalaya–Hindu Kush region, which includes parts of China and India and contains more ice than anywhere else on earth except for the polar ice caps. The portion of Peru covered by glaciers has shrunk by 25 percent in the past 30 years. Glaciers in Norway, France, and Spain are also disappearing at a rapid rate. Most of the 23 trillion gallons of water used by Californians each year comes from Sierra Nevada snowmelt, which is decreasing each year.

Rising temperatures will produce more rain and less snow in the mountains, and hence rivers and streams will carry more water during the rainy season and there will be a shortage of water at other times. Another effect of global warming is droughts that last for extended periods of time. Lack of rain over the last few years has made Australia "the parched continent."[10] The local climate turned to the other extreme in 2010, the wettest year in recorded history, with devastating floods in many parts of the country. The flow of water in many rivers around the world has decreased due to regional drought conditions. From 2000 to 2007, the Colorado River experienced the worst drought in the last 100 years of recorded history; the water stored in Colorado River reservoirs has dropped from nearly full to less than 55 percent of capacity.[11] At the end of

July 2008, most parts of the southeastern United States were classified as in moderate to severe drought, presumably caused by climate change associated with global warming.

GROUNDWATER

Because seasonal rains often do not provide enough water to grow grains and other crops, and freshwater from rivers, streams, and lakes in many parts of the world is not easily available or has been fully exploited, the agricultural output of many regions is heavily dependent on groundwater held in aquifers. From ancient times, farmers have dug into the ground to obtain additional quantities of water for irrigation.

There are thousands of aquifers all over the world that have stored water over eons. The total amount of groundwater is much greater than all other sources of freshwater, excluding the amount that is frozen in ice caps and glaciers. There are basically two types of aquifers: the shallow alluvial aquifers that are recharged through rainfall, and the so-called fossil aquifers that accumulated water in the ancient past when the local climate was much different, with either abundant rainfall or a fair amount of snow. Most of the aquifers in India and the shallow aquifer under the North China Plain are rechargeable. When rechargeable aquifers are depleted, the maximum rate of pumping is automatically reduced to the rate of recharge, usually determined by local rainfall.

The three largest fossil aquifers are the vast Ogallala Aquifer in the United States, the deep aquifer under the North China Plain, and the Saudi Arabian aquifer. Although the amount of water stored in aquifers all over the world is very large, it can be exhausted if the rate of replenishment is less than the rate of withdrawal. Fossil aquifers are located in regions with small amounts of rainfall; thus, they are not significantly recharged. Water stored in the fossil aquifers is a precious resource that we have inherited from previous ages, somewhat like fossil fuels. For this reason, removal of water from these aquifers is called water mining.

Groundwater is used in many parts of the world these days to compensate for deficits in freshwater. Powerful diesel and electrical pumps allow farmers to dig deeper and deeper to extract water that is required to grow crops. Using advanced technology, millions of wells have been dug in all parts of the world. A lowering of the level of groundwater is a clear indication that groundwater is being withdrawn at a rate greater than the rate of recharge. While a small seasonal decline is acceptable, a progressive decline that continues for years on end represents a serious problem. Modern multi-stage pumps have a much greater reach than earlier versions; still, if the level of water keeps falling, the cost of bringing water to the surface will eventually become prohibitive.

Water tables in all the continents are in precipitous decline because water has been used at rates that are much greater than the rates of recharge. Urbanization has greatly contributed to this decline as well, as buildings and paved surfaces prevent rainwater from infiltrating the soil, and millions of gallons of rainwater are lost through drains and rivers. Groundwater depletion is most severe in parts of Asia, the United States, North Africa, and the Middle East. Water tables are falling at the rate of 3 to 10 feet (1 to 3 meters) each year in many countries, including the three largest grain-producing countries: India, China, and the United States. In India, which leads all other countries in irrigated area, water for half the country's croplands is extracted from underground reserves. About 21 million wells have been drilled in that country, mostly with hand pumps, and they are lowering the level of groundwater almost everywhere.[12] In North Gujarat, the water table is falling at rates that may approach 20 feet per year, and in the southern state of Tamil Nadu, 95 percent of the wells on land owned by small farmers have dried up.

Groundwater plays a crucial role in the agriculture of the United States and China. The area of irrigated farmlands in the United States is the third largest in the world; 40 percent of these farmlands depend on groundwater for irrigation. The agricultural region in the North China Plain, which produces 40 percent of China's grain, is primarily irrigated with water from the fossil aquifer. Due to falling water levels, Chinese wheat farmers often have to dig wells to a depth of a few hundred feet to get the water they need. When water falls to very low levels, the cost of pumping becomes so high that farmers stop irrigating the fields. The level of groundwater is now even lower in some parts of Saudi Arabia, with reports of wells as deep as 4,000 feet.[13] Many countries, including Pakistan, Iran, Yemen, and Spain, are seeing a rapid lowering of the level of groundwater. Out of the 188 most important aquifers in Mexico, 80 have been overexploited to various degrees, including some that are severely overexploited.[14]

The Ogallala Aquifer

The Ogallala Aquifer, also known as the High Plains Aquifer, is the world's largest aquifer system and supplies water to most of the irrigated farmlands in the central United States. An extensive underground reservoir, it covers an area of 175,000 square miles (450,000 sq km) in parts of eight states: South Dakota, Nebraska, Wyoming, Colorado, Kansas, Oklahoma, New Mexico, and Texas. Water accumulated in this aquifer over 20 million years ago when gravel and sand from the Rocky Mountains were eroded by rain and washed downstream. Those sediments soaked up water from rain and melted snow and preserved it in the aquifer for millions of years in spongelike structures. Some water from Ogallala provides for the needs of local communities for household activities and industrial operations, but the bulk of it is used to irrigate farms that produce a large portion of the grain in the United States. It has fostered the development of large-scale agriculture in the arid part of the country and has redefined the landscape of the central United States by promoting economic expansion and population growth.

The invention of powerful pumps in the late 1950s spurred the development of a method of irrigation known as the center-pivot system that is used extensively to irrigate farmlands with groundwater from Ogallala. The center-pivot system extracts water from deep wells with high-power pumps and sprays it with a rotating arm long enough to cover a 160-acre field with each sweep. This method of irrigation is responsible for the characteristic circular patterns visible from the air in most agricultural landscapes of the western United States. Due primarily to this invention, the acreage of cropland irrigated from the High Plains Aquifer increased from less than 2.1 million acres (850,000 hectares) in 1949 to 13.7 million acres (5.6 million hectares) in 1975, and the amount of groundwater withdrawal increased from 4 million acre-feet (4.9 billion cubic meters) in 1949 to 19 million acre-feet (23.4 billion cubic meters) in 1975, reaching 21 million acre-feet (25.9 billion cubic meters) in 2000.[15] Vast quantities of corn, alfalfa, soybeans, and wheat—grown with the help of water from Ogallala in Nebraska, Kansas, Texas, and other states—provide cattle operations with the feed required to produce 40 percent of the feedlot beef in the country.

The Ogallala Aquifer is being exhausted because its water is being removed at a rate greater than the rate of replenishment. In fact, the rate of replenishment is minimal by comparison. The U.N. Environment Programme estimates that the aquifer has been depleted of at least one-fifth of its stored water, and some experts have declared it to be the most rapidly disappearing aquifer in the world. Many wells in Texas and Kansas have gone dry, and many more are in danger of being exhausted. According to the Kansas Geological Survey, the water level of the Ogallala Aquifer has fallen by an average of 5.47 feet, with a decline of up to 30 feet in some places.[16] The depth of the water table in the aquifer has also declined from predevelopment values in many regions; 13 percent of the aquifer area has sustained more than a 25 percent decrease, and 5 percent of the aquifer has more than a 50 percent decrease.[17] Neither agricultural activity nor water requirements are uniformly distributed throughout the region, which means that the areas of intense agricultural activity have lost the greatest amounts of water. Depletion of groundwater in the aquifers is already having a deleterious impact on agricultural activity in these regions, and loss of crops will increase with the passage of time as the wells continue to dry up.

Irrigating farms with giant sprayers is a wasteful practice, because large amounts of water evaporate and never reach the plants. Center-pivot irrigation pumps remove water at the rate of 750 gallons per minute. Some wells in Texas and New Mexico are so depleted that they do not even have enough water to operate these pumps; these farmlands are being abandoned, and farm foreclosures are not uncommon there. Looking back to the middle of the last century, we might recall that the semiarid High Plains were plagued by crop failures due to cycles of drought. These failures culminated in the disastrous Dust Bowls of the 1930s, graphically described by John Steinbeck in his novel *The Grapes of Wrath*.

Before the Ogallala Aquifer was discovered, much of the land in the southwestern United States was considered suitable only for grazing cattle, not for intensive farming. Water from the Ogallala Aquifer transformed the land into one of the most productive regions of the world and has redefined the landscape of this large area. Without water from this aquifer, vast stretches of land would revert to their original desertlike state. Since the United States is a major exporter of agricultural products, this change would have a serious impact on the availability of food everywhere.

On almost every continent, major aquifers are being drained faster than their rate of recharge. Groundwater depletion is most severe in parts of India, China, the United States, North Africa, and the Middle East. In addition to the regions covered by the Ogallala Aquifer in the United States, the Great Plains and the Southwest are seeing falling water tables. In India, they are falling in most states, including Punjab, the nation's breadbasket. In China, water tables are falling throughout the northern half of the country, including under the North China Plain, the source of half of the nation's wheat and a third of its corn.

Groundwater in aquifers is a precious resource that we have inherited from previous ages. Shortages of water already limit how much food can be produced. If we exhaust all sources of freshwater from the surface and levels of groundwater continue to decline, the situation will progressively become worse. By using water from fossil aquifers at a rapid rate, we are also limiting the options available to future generations.

RIVER FLOW, POLLUTION, AND OTHER CONSIDERATIONS

Groundwater is also the source of the water flowing in many rivers and streams. If the level of groundwater goes down, there will be less water available for rivers, which will affect their flow. While the Mississippi, the Yangtze, the Niger, and many other major rivers of the world receive a substantial contribution from groundwater, many smaller rivers are entirely dependent on groundwater to maintain their flow. Because of declining levels of water, the state of Kansas alone has already lost more than 700 miles of rivers that once flowed year-round. A decrease in the flow of water anywhere harms plant and animal life that depends on it, but the effects are especially devastating in semiarid or arid regions, because all forms of life in these areas are more dependent on water.

Groundwater is usually of good quality because it is filtered through layers of soil, and its rate of accumulation is very slow; however, once it is polluted with chemicals or other substances, it is very difficult to reverse the process to reestablish its purity. Intense agricultural activities using typical modern farming methods invariably pollute groundwater with fertilizers, insecticides, herbicides, and other chemicals. This pollution does not remain localized where these chemicals are applied but is distributed over a large area, including neighboring bodies of water. Because this type of pollution builds up slowly, it may go undetected for many years. The problem of pollution worsens as the level of water falls, because more agricultural chemicals leach into proportionally smaller bodies of water. Since we cannot see groundwater, when it is depleted or contaminated it does not have an emotional, visible impact on us, as the pollution of a river or the drying up of a lake or stream does.

Rapid depletion of groundwater is likely to create other problems. Removing water from the soil creates a void that may allow seawater to intrude—a phenomenon that occurs usually, but not exclusively, in coastal regions. A mixture that is just 6 percent saline water makes groundwater useless for human consumption or agricultural use. Saline intrusion has already degraded the productivity of land in coastal cities such as Bangkok, Dhaka, Jakarta, Karachi, and Manila.

A related problem caused by emptying the land of its water content is subsidence—either a gradual settling or a sudden sinking of the earth's surface. Subsidence occurs worldwide, primarily by compaction of the layers of soil due to removal of groundwater. About one-quarter of the weight of the upper-level soil consists of moisture; removing this moisture naturally affects the integrity of the ground. The type of soil and whether there are underground rock formations or not influence how rapidly and drastically land sinks.

Land subsidence and fracturing are problems in many parts of the world, including the Las Vegas Valley, Mexico City, Shanghai, and Long Beach. In the San Joaquin Valley in central California, almost 30 percent of the land surface has sunk below the original level. According to the U.S. Geological Service, 17,000 square miles (44,000 sq km) in the United States—an area roughly the size of New Hampshire and Vermont combined—has been directly affected by subsidence.[18] This trend will accelerate as groundwater is depleted in other parts of the world.

Tanning the hides of animals to make consumer goods both uses and pollutes water. Each year approximately 5.5 million tons of raw hides are processed to produce more than a million tons of finished products.[19] The wastewater from tanneries contains preservatives and chemicals such as chromium salts, fungicides, bactericides, and solvents that are injurious to all forms of life. These components adversely affect water quality and damage the ecosystem, causing harm to fish and other aquatic life. The amount of polluted water is small to begin with, but it mixes with neighboring bodies of water and degrades the quality of water in a much larger area if pollutants are concentrated.

FOOD FOR THE FUTURE

Overexploitation and pollution of all sources of freshwater, coupled with the situations created by climate change accompanying global warming, will severely constrain how much food will be available during the coming decades. As many as a billion people today are eating food grown using underground water that is not being replaced. The pressures of increasing population and the demand for animal products have brought the limit of this exhaustible resource within sight, with the distinct possibility that the ecosystem may not be able to provide food for everyone far into the future. Choosing grains, fruits, and vegetables rather than animal-based foods can extend the time for which food will be available in sufficient quantities for the earth's population.

The coupling of food habits and exhaustion of resources is best illustrated by China, a country that had always produced enough food for its people. According to statistics maintained by the Food and Agriculture Organization of the United Nations, the average meat consumption in China was 44 pounds (20 kg) per person annually in 1986. By the year 2000, it had increased to 110 pounds (50 kg), and China was forced to import a lot of food. Since markets are now connected at the international level, if China continues to import large amounts of food, there will be serious repercussions for the supply of food almost everywhere.[20]

A shortage of water may make it necessary for all countries to examine how much water is needed to produce each item—to take the so-called virtual water into account. Since the shortage of water is not being taken seriously in many parts of the world, the market price of goods does not reflect their virtual water content. At present, Canada exports its water in the form of grains, Brazil and Argentina in the form of beef, and Thailand in the form of rice. Even though projections by the U.S. Geological Survey point to an impending shortage of water in the United States, about a third of the water that is withdrawn from the natural environment is exported in the form of agricultural items.[21] Many countries in northern Europe and Southeast Asia depend on importing virtual water from the Americas, Africa, and Australia, even though some of these lands do not have excess water.[22]

The impending shortage of water in many parts of the world is a dark cloud on our horizon. Sustaining future generations will require a multiplicity of approaches, including conservation of water for all uses, irrigation methods that are targeted to deliver water to plants without wastage, and horticultural research to develop varieties of plants that can tolerate limited periods of drought. Considering that agriculture accounts for more than 70 percent of the total water used in all human activities, however, we can see that its requirements are too large to be met by minor changes.

The global population is projected to be 7.9 billion by the year 2025 and 9.1 billion by 2050. Making food available to the vast majority of people in the world will depend on the availability and judicious use of water. An extrapolation of the present rate of consumption shows that the developing countries will need 50 percent more water by the year 2025, while the requirement of developed countries will increase by 18 percent[23]—and the water is simply not there. While the need is to procure greater amounts of water to meet future needs, it appears unlikely that the water sources will continue to supply at the present level for long. Food is an international commodity; grains, meats, fruits, and other food items frequently cross national boundaries, even continents. A shortage of food in some parts of the world is bound to have repercussions everywhere.

* * *

Drastically reducing the consumption of animal-based foods will have an immediate and far-reaching effect on preserving the supply of water, because, as I have stressed, the livestock sector consumes prodigious amounts of water, both directly and indirectly. Raising livestock for food is depleting and degrading all types of freshwater sources—rivers, streams, lakes, and groundwater. Human populations and demand for meat are both accelerating the process at a rapid rate. We are consuming water that belongs to future generations and living on borrowed time by mining the aquifers—dipping into the slowly renewable or nonrenewable water cycle. As the environmental reporter Fred Pierce has succinctly phrased the problem, "We will live to regret this, and if we don't, our children will."[24]

5 * FISH

The creatures who inhabit the earth's seas, oceans, rivers, and lakes have provided food for humans since antiquity. Water covers more than 70 percent of the earth's surface—but the productivity of seas and oceans is limited by the fact that only about 0.03 percent of the energy from the sun is captured and converted to a useful form by phytoplankton (which includes algae, cyanobacteria, and some other microscopic creatures capable of harvesting the energy of the sun through photosynthesis).[1] This is why the amount of food people glean from marine life is equivalent to less than 1 percent of the amount of food we eat that comes from the land. The yearly consumption of fish and shellfish in the United States is 47 pounds (21.3 kg) per person, somewhat greater than the global average of 35.5 pounds (16.1 kg). The amount of fish eaten in different parts of the world depends on people's proximity to the sea.

DECLINING STOCKS OF WILD FISH

People living in coastal communities have always fished as a source of nutrition and income. Before the introduction of large vessels and trawlers in the 1950s and 1960s, fishing was limited to coastal regions, and the ocean was able to replenish our harvests of marine life. The new large vessels, however, travel thousands of miles and greatly increase the haul of wild fish. More recently, even larger ships and sophisticated radar, sonar-echoing devices, and global positioning systems allow every part of the ocean to be exploited by modern fishing fleets. And yet the total weight of wild fish caught has been declining since 1994 at the rate of 700,000 tons per year, because overfishing is decreasing the net productivity of the oceans.[2] Using a meta-analytical approach, Daniel Pauly and colleagues at the University of British Columbia have estimated that the large predatory biomass of fish and seafood today is at only about 10 percent of preindustrial levels.[3]

An example of the decreasing stocks of wild fish is provided by the decline of salmon fishing on the west coast of the United States. Salmon is the third most popular seafood in the United States, after shrimp and tuna. California fishermen bring in about five million pounds of meat of the highly prized chinook, also known as king salmon, in good years. The fish return to the Sacramento River to spawn. Young fish swim downstream and live in the Pacific Ocean for the next three years, until their body mass increases to 10 to 50 pounds, at which time they retrace their parents' path to spawn in the river. In the early phase of their lives, when the young fish are in the ocean, they ingest the tiny shrimplike creatures known as krill that grow in huge numbers in cold waters, imparting a reddish color to the sea. It is the diet of krill that gives a pink color to the salmon.

Although there have long been significant fluctuations in the number of returning adult salmon, for years they consistently numbered more than 500,000, and they numbered 750,000 in 2002. In 2007, however, their number dropped precipitously to 90,000, and in 2008, they declined to 66,200. According to the Pacific Fisheries Management Council (PFMC), a returning population of 120,000 to 180,000 adults is required to prevent a collapse of the wild salmon in that area. In 2009, the decision to ban commercial fishing from the coasts of California to north-central Oregon for the second year in a row was necessary if there were to be any hope of salmon recovery.[4] Even after a two-year ban, the number of returning salmon was predicted to be only 245,000 in 2010 when the PFMC lifted the ban on commercial fishing for two four-day periods.[5] Sadly, a recovery is becoming more unlikely with each passing year. The situation is equally dire for other important sources of wild salmon, such as those in the Yukon River in Alaska.

The loss of marine life and declining productivity of the seas can be established from a number of observations and expressed in many ways. Marine biologists define a particular species of fish to be collapsed if its catch is less than 10 percent of the recorded maximum. Using this criterion, scientists conclude that 29 percent of the currently fished species have collapsed, and many more are in danger of collapsing.[6] At the current rate of harvesting, all edible fish species will collapse around the middle of the twenty-first century, with serious consequences for the availability of seafood everywhere. Only extreme actions can prevent the harvest of fishes and other seafood from drastically decreasing within just a few decades. Such actions are unlikely because of overriding regional interests and the lack of an enforcing authority in the open seas.

Another way of looking at the effect of large-scale commercial fishing on marine life is to examine the so-called trophic levels (TLs) of the species that have been caught. Trophic level indicates the position of an organism in the food chain; those with a higher trophic level survive by consuming species at a lower trophic level. Algae, at the bottom of the food web, have a TL of 1, and herbivorous zooplankton that feed on algae have a TL value of 2. Large fishes, whose food tends to be a mixture of low and high TL organisms, are assigned a TL value between 3.5 and 4.5. Fishery operations tend to remove larger, slower-growing fishes, thus reducing the mean trophic level of the fish remaining in the waters. Data collected by marine biologists show that the trophic level of the wild fish harvested from oceans is gradually decreasing, indicating that we are gradually "fishing down the marine food chain."[7] Larger species that live longer are slowly being removed from the world's oceans.

Fishing with Trawlers

In response to the depleted stock of fishes, coupled with the increasing demand for seafood, the fishing industry has adopted practices that are accelerating the collapse of edible fishes by destroying the habitats of marine organisms and damaging the chain of life that is essential for the long-term survival of many species. Most commercial fishing is done with fishing trawlers, large and powerful boats that drag a net through the ocean as sophisticated locating devices guide them to schools of fish. There is no part of the oceans that is not being explored by large trawlers in search of fish. The increasing capacity, accuracy, and range of fishing fleets are the main threats to the long-term survival of fisheries in seas and oceans.

While trawler fishing is always done with nets, using the method known as bottom trawling, large ships drag huge, heavy nets along the bottom of the ocean for thousands of miles. The bottom of the ocean is a complex ecosystem in which creatures of various kinds develop suitable habitats around available mud, grass, coral reefs, and rocks. Numerous types of fishes and invertebrates utilize these features for laying their eggs, catching prey, and hiding from predators. The nets of bottom trawlers capture, crush, uproot, or destroy everything in their paths. Reestablishing marine life around a landscape obliterated by bottom trawlers may take a very long time, because species are interdependent. Elliott Norse and Les Watling of the Marine Conservation Biology Institute (MCBI) have calculated that tens of thousands of trawlers working in the oceans scrape nearly six million square miles, twice the area of the lower 48 U.S. states, each year.[8] They operate in almost all regions of the world, including continental shelves and deep oceans, destroying marine life on a massive scale.

The nets of trawlers invariably capture large quantities of fish belonging to the nontarget species, known as trash fish or bycatch. Since fishing vessels have limited capacity for freezing and canning, they are focused on the target species, and they simply discard any bycatch. According to the scientists of the MCBI, a trawler in search of shrimp and finfish hauls 20 pounds of other species for every pound of the fish they seek; all trawlers put together discarded bycatch estimated to weigh about 30 million tons.[9] This figure is likely to be a low estimate, because bycatch figures are underreported, and statistics do not include fish lost to spoilage. For trawler operations in the Gulf of Mexico, 16 percent of the total catch was commercially valuable shrimp, while 68 percent was unintended bycatch, mostly of juvenile finfish.

In addition to depleting marine stock almost everywhere, these large ships represent a bad bargain from purely energy considerations. It has been estimated that trawler fishing requires 3.4 liters of fuel oil per kilogram (or about 0.4 gal per lb) of fish; the input of fossil fuel for catching fish is roughly 14 times larger than for the production of vegetable proteins.[10] Because their large consumption of fossil fuels makes fishing trawlers economically nonviable, this kind of fishing is sustained by subsidies from the governments of many countries. These subsidies, which account for about 20 percent of the value of the fisheries sector, have created excess fishing capacity, which is outstripping available fishing resources. The global fishing fleet is approximately 250 percent greater than needed to maintain the catch at a sustainable level.[11]

Pollutants in the Waters

Human activities such as dumping waste products and industrial effluents have been adversely affecting the productivity of the seas for centuries. Many people have seen the oceans as having an inexhaustible capacity and as being too large to be adversely affected by human endeavors. But many pollutants are so dangerous that they either kill aquatic creatures or render them toxic and unfit for human consumption. Industrial pollutants that are somewhat less potent stay in the food chain of marine creatures and are eventually ingested by humans who eat seafood. For example, mercury and heavy metals are deposited in high concentrations in fishes and mollusks. Some species of fish are found to contain significant levels of methylmercury, PCBs, dioxin, and other environmental contaminants—so fish and other seafood are a major source of exposure to these substances. PCBs and methylmercury have long half-lives in the human body, and they accumulate, damaging the organs of people who consume large amounts of contaminated fish.

Of all the toxic substances in fish, mercury has received the greatest attention, because it is present at various levels in all types of fish and other aquatic organisms. Over the years, many companies have used mercury to manufacture a range of products such as thermometers, thermostats, and light switches. It is released into the air through the smokestacks of factories as industrial pollution and eventually finds its way into the sea through rivers and streams. Once mercury enters a waterway, naturally occurring bacteria absorb it and convert it to a form called methylmercury. This transition is particularly significant for humans, who absorb methylmercury easily and are especially vulnerable to its effects. Once in the human body, mercury acts as a neurotoxin, interfering with the brain and nervous system. The recent trend of replacing incandescent light bulbs with fluorescent bulbs to save energy increases the industrial use of mercury, because these devices contain traces of mercury. The species of fish that usually have the highest levels of mercury include tuna, swordfish, sea bass, and bluefish.

Dead Zones in Coastal Regions

Overfishing and pollution are having a great impact on the fish population. The discharge of nutrients into coastal waters is decreasing their fecundity and decimating the marine organisms in some regions. The continental shelf is the portion of the seas that is closer to the land and is much shallower than the open oceans. It extends into the sea, on average, 73 kilometers (about 45 miles) and varies in width from negligible amounts to 1500 kilometers (about 930 miles). Its average depth is a few hundred meters. The physical and chemical characteristics of the continental shelf regions are more variable than those of the open ocean, since they are influenced by seasonal variations in the land. The combination of ample nutrients and shallow waters creates a favorable environment for aquatic life, so a great variety of marine organisms thrive near the continental shelves. The surface area of these regions is only 10 percent of the total area of seas and oceans, yet they provide more than 90 percent of the world's harvest of fish and shellfish. Compared to the teeming life and high productivity near the continental shelves, the oceans are like deserts inhabited by larger fish and relatively few species of mammals.

Marine life suffers large losses when nutrients such as untreated human waste, discarded biological materials, and fertilizers are dumped into the coastal seas. Only a fraction of the synthetic fertilizer applied to farmlands is absorbed by the roots of the crops; the rest is washed away by irrigation, with a substantial portion finding its way to the sea through rivers and streams. Fertilizers and biological waste trigger the bloom of specific types of algae. Some of the algae that proliferate in these regions produce toxins that kill fish or contaminate them so humans cannot safely eat them. Even algae that are not toxic do damage to aquatic life in these regions. As the algae rot and die, they remove dissolved oxygen from the water. Since the oxygen content of water is a crucial factor for sustaining all forms of life, bodies of water with excessive levels of nitrogen and nutrients become "dead zones" that cannot support higher organisms such as fish, clams, lobsters, and oysters.

Hypoxia (a deficiency of oxygen) and anoxia (extreme hypoxia) are becoming more and more frequent worldwide; they cause massive kills of invertebrates and fish. According to the U.N. Environment Programme's *Global Environmental Outlook Year Book,* in 2003 there were 150 dead zones around the globe—double the number in 1990. A more recent study indicates that dead zones have spread exponentially; 400 such zones were observed in 2007, encompassing a total area of more than 245,000 square kilometers (about 95,000 sq miles).[12] Such zones exist in the Gulf of Mexico, the Chesapeake Bay, the Baltic Sea, the Adriatic Sea, the Gulf of Thailand, and near the continental shelves in South America, Japan, China, Australia, New Zealand, and other parts of the world. Some of these hypoxic zones cover tens of thousands of miles, and their borders keep shifting, so living creatures in neighboring areas are suffocated, too.

There is a connection between the dead zones and the production of animal-based foods. According to the U.S. Geological Survey, the hypoxic zone in the Gulf of Mexico, where the Mississippi River discharges water from the agricultural fields of midwestern states, covered an area of 20,000 square kilometers (about 7,722 sq miles) in 2006, roughly the area of the state of New Jersey.[13] The nutrients carried by the Mississippi River are primarily residues of fertilizers applied to farmlands in the Midwest that grow feed for the livestock industry. Excessive application of fertilizers to produce grains for the livestock industry is responsible for many dead zones all over the world. Since these dead zones usually occur near the highly productive continental shelves where many varieties of fish spend the early part of their lives, they have a huge negative impact on marine life.

AQUACULTURE

Fish farming, known as aquaculture, involves raising one species of fish, often genetically engineered, in enclosed bodies of water under controlled conditions. Salmon, carp, catfish, and tilapia, as well as invertebrates such as shrimp, clams, mussels, and oysters, are raised on fish farms. Far Eastern countries produce the bulk of fish raised in farms: China is the largest aquaculture producer, accounting for about 70 percent of the world's production, and 80 percent of the shrimp sold in the United States originates from aquaculture farms in Asia, primarily Thailand, Vietnam, and Cambodia. However, fish farms are now increasing in number in many countries in Europe and the Americas. Aquaculture facilities in Latin America primarily cater to the needs of the developed world. Worldwide, the aquaculture industry has been growing at a phenomenal rate. The output of fish farms increased from 5 million tons in 1980 to 36 million tons in the year 2000. The contribution of aquaculture to total fish production was less than 4 percent in 1970 but increased to almost 29 percent in 2002 and 36 percent in 2006, representing an annual growth rate of 7.2 percent,[14] three times greater than for land-based animal production.

Table 5.1. World fisheries and aquaculture production and utilization, in millions of tons

	Wild	*Aquaculture*	*Total*
Inland	9.6	28.9	38.5
Marine	84.2	18.9	103.1
Human consumption	—	—	107.2
Nonfood use	—	—	34.4

Source: World Resources Institute, http://earthtrends.wri.org/.

In a typical aquaculture facility, fish eggs that have been genetically selected or modified are raised in a hatchery, and then the young fish are transferred to either inland ponds or sections of the sea separated from the main body of water by nets. Farms in Asian countries often use inland ponds close to the sea to raise shrimp, tilapia, and fishes, whereas they demarcate an area within the sea with nets to raise salmon. Some of the species raised in aquafarms, such as tilapia, oysters, and carp, are herbivores or omnivores, while others, including shrimp, salmon, and bass, are carnivores that require fish products in their diet. The success of fish farming often depends on the availability of nutritionally suitable feed. Fish farming also requires copious amounts of water. For fish raised in ponds, the water has to be changed on a regular basis to prevent waste materials from accumulating to toxic levels; cages in coastal areas pollute the local environment to such an extent that they have to be relocated every few years.

To increase production and profits, the density of fish in these enclosures is kept very high—the maximum stocking density is limited only by excessive mortality in the population. Salmon farms usually raise as many as 50,000 individuals in each enclosure and are so crowded that a 2.5-foot-long fish spends its entire life in a space the size of a bathtub. Trout farms are even more crowded, with up to 27 full-grown fish in a bathtub-size space. Such crowded conditions are conducive to the growth of microscopic pathogenic organisms that grow at very low levels, or not at all, in the wild. These bacteria and viruses sometimes reach epidemic proportions and wipe out the whole fish colony. The pathogens can invade fishes' bodies through the gills and, unlike with farm animals, signs of sickness in individual fish are not easily spotted by the farms' operators. This delay in recognizing trouble means that the spread of disease becomes obvious only when a large number of fish begin to die.

A common problem in fish farms is outbreaks of sea lice, which eat the outside of the fish, causing their scales to fall off and creating large sores. In severely crowded conditions, sea lice eat down to the bone of the fish's faces. This situation is so common that fish farmers call it the "death crown." Densely packed fish also rub against each other and the sides of the cages, damaging their fins and causing infections. Aquaculture operators use strong antibiotics to control sicknesses and diseases in their stocks, but mortality rates of 30 percent are not uncommon. Aquaculture facilities in China's Jiangsu province reported mortality rates of 20 percent for finfish, 40 percent for shrimp, and 50 percent for mollusks.[15] There have been instances when the entire shrimp population of a farm has been wiped out by a viral disease.

In many respects, aquaculture is similar to raising livestock in concentrated feeding operations, because the marine creatures pollute the local environment and require input from distant places that eventually bear the scars of their impact. Even under optimum conditions, a significant part of the feed is not eaten by the fish and ends up on the ocean floor. The area around the aquaculture farms is polluted with tons of fish feces, antibiotic-laden fish feed, and diseased fish carcasses. Accumulation of these substances sometimes causes the ocean floor around these facilities to rot, eliminating all useful forms of life from those regions. Anaerobic fermentation of uneaten feed and feces generates hydrogen sulfide, which can impair fish health and cause sudden mass infertility. Aquaculture facilities that are only separated from the sea by nets continuously exchange water with neighboring regions. Operations that raise fish in ponds have to discharge their water into the sea every few days to keep the pollution at an acceptable level. Hence the drugs administered to the fish, as well as the pathogens that afflict the aquaculture stock, end up in the sea, damaging the wild fish in the region.

Shrimp, tuna, and salmon are the three most highly prized types of seafood in the developed world. Shrimp is often obtained through aquaculture, mostly in the Far Eastern countries. The most suitable location for such facilities is in mangroves—the lush, forested buffer zone between the land and the sea. Mangrove forests are vital as breeding grounds for many species of commercially important wild fish. These ecologically sensitive and productive regions are being eliminated at a rapid rate to make room for shrimp farms—at least 60 percent of the original forests have already been lost. Shrimp farming is often done in a predatory manner; the ponds are abandoned when the level of pollution created by them becomes too high to sustain their operation, leaving devastated coastal habitats and communities in their wake. Most aquaculture farms need a substantial input of freshwater to reduce the concentration of undesirable materials and to maintain salinity at a tolerable level. The requirement for freshwater is particularly high for shrimp farms; it takes 50,000 to 60,000 tons of freshwater to raise a ton of shrimp.[16] In ponds with intensive production, about a third of the water has to be changed daily, making this industry a heavy consumer and polluter of water. The use of antibiotics in aquaculture facilities leads to the development of resistant strains of bacteria in farmed fish and also in the fish and shellfish harvested from areas surrounding these enclosures.

Salmon farming pollutes the local environment and probably causes more problems per pound of fish than any other type of aquaculture. Millions of salmon escape from their cages each year, disrupting wild salmon populations by introducing diseases and parasites and competing with them for habitat. Faster-maturing and more aggressive farm salmon may initially deprive the smaller, more cautious, wild fish of food and shelter, but ultimately the farm fish fail to survive because they are bred to live in controlled environments and do not have the hardiness of wild varieties. Mating between the two groups introduces these vulnerabilities into the wild stock, further endangering their survival.

Many species that are raised in aquaculture farms, particularly those that are highly prized, like salmon, bass, trout, and shrimp, are carnivores that require fishmeal and fish oil and cannot survive on an herbivorous diet. These fish food items are byproducts of trawlers that indiscriminately haul all kinds of aquatic life into their nets. After separating and processing the desired species as food for humans, everything else is ground up and becomes fishmeal for aquaculture and livestock farms. Fishmeal and fish oils are the only available sources of highly unsaturated fatty acids that are both essential nutrients for all carnivorous fish and a key reason for their health appeal. It takes three to six pounds of wild fish to produce one pound of farmed fish. Around 11 million tons of wild fish—

about 12 percent of the total haul from seas and rivers—are fed to farmed fish.[17] Fishmeal used to fatten high-value salmon or bass often includes herring, sardines, and mackerel that could be consumed by people.

All carnivorous fish and shrimp species raised in aquafarms are net protein consumers rather than protein producers. Attempts to raise carnivorous fish on a purely vegetarian diet have been mostly unsuccessful, because the fish suffer from irritated lower intestines and a depressed immune system and lose their ability to absorb key minerals such as zinc and iron. Through detailed analysis, Naylor et al. have shown that aquaculture cannot compensate for the loss of wild fish and that these operations depend on the already severely depleted stock of marine animals.[18] Salmon, shrimp, and trout aquaculture use up almost 50 percent of all fishmeal used in aquaculture but provide less than 10 percent of the total production volume.

Aquaculture is hastening the process of depletion of wild stocks of life in the open seas. In search of feed for farm-raised fish, some operators have begun harvesting krill, the tiny shrimp-like crustaceans that are found in cold waters. These organisms feed on phytoplankton and are near the bottom of the food chain. They form an important component of the diets of larger fish and other animals. Large-scale harvesting of krill is bound to have an adverse effect on the survival of all forms of marine life and thus is a process that sacrifices sustainability in favor of short-term gains.

The level of beneficial omega-3 fatty acids varies significantly in farmed Atlantic salmon (the species that is generally raised in aquafarms), depending on the type of feed given to them. In some but not all cases, salmon from aquaculture operations have been found to have low levels of omega-3 fatty acids and high levels of omega-6 fatty acids, which reduce the health benefits of eating fish. Using vegetable oils instead of fish oils in the fish feed produces milder-flavored, less oily fish filets with a different texture. Farmed catfish, tilapia, and shrimp are very low in fat and therefore low in omega-3 fatty acids, although they are good sources of protein. A vegetarian diet fed to fish that can survive on such feed tends to produce lower amounts of omega-3 fatty acids in their flesh, thus reducing their beneficial effect on human health. A number of studies have also found surprisingly high levels of toxic chemicals such as PCBs and dioxins in farmed salmon.[19]

While the stock of wild fishes is decreasing and their habitats are being destroyed, the demand for fish as food is increasing, primarily because of its presumed health-related benefits. It is often suggested that the declining productivity of the oceans can be more than compensated for by the use of aquaculture. Although aquaculture seems to be filling the gap in the harvest of wild fish created by their diminishing stocks, there are ample reasons to believe that fish farming cannot be sustained at increasing levels for long. As in meat production, in fish farming feed is a major bottleneck. It is evident that the growth in aquaculture may be limited by access to feed, which in turn is partly dependent on capture fisheries. In fact, the growth of aquaculture production has already slowed. The yearly growth rate was 11.8 percent in 1985 but only 6.1 percent from 2004 to 2006.[20] Even if the increasing demand for fish is ignored, maintaining per capita fish consumption at present levels will require a continuous increase in the catch of wild fish that cannot be sustained. A future collapse of ocean fisheries, as has been projected by a number of studies, would immediately affect aquaculture production.

We know enough about ocean processes to conclude that marine resources cannot keep pace with the ever-increasing demand for fish. Preserving the existing stocks of marine life would require imposing an embargo on wild fish, a course that is neither practical nor palatable to a host of communities that would lose their incomes and livelihoods. In the absence of such actions, we are inexorably sliding toward a situation where the productivity of the seas will be extremely limited.

Aquaculture operations have other undesirable effects. Fish farming requires both land and water, two resources already in short supply in many areas. In Thailand, both of these resources have been used in recent years to fuel the growth of the aquaculture industry; nearly half the land now used for shrimp ponds was formerly used for rice paddies. In addition, water diversion for shrimp ponds has lowered the level of groundwater in many coastal areas. Industrial production of shrimp is carried out mostly in developing countries in tropical regions. In many of these nations, mangrove forests and wetlands have been destroyed to make way for fish farms. Aquaculture of carnivorous fish is commercially viable, even though it uses much more fish than it produces, because it is directed toward wealthy customers in developed countries. The fish produced is beyond the means of the local population. By occupying precious territory in coastal regions, aquaculture often deprives the local community of fishing opportunities. Factory farming, both in water and on land, allows control of the industry by large multinational corporations. A handful of companies produce two-thirds of the farmed salmon and trout in various regions of the world.[21]

LONG-TERM IMPACT

The loss of populations and species of aquatic creatures in human-dominated marine ecosystems, which now means all bodies of water, is accelerating. Overexploitation by fisheries is leading to the gradual elimination of large, long-lived fishes from marine ecosystems and their replacement by shorter-lived fishes and invertebrates operating within food webs that are much simplified and lack their former buffering capacity. Since the web of marine life is highly interdependent, the loss or severe decline of one species may adversely affect the productivity and stability of many other species. A complex aquatic system with a great variety of organisms is more resilient than one with fewer species, because the collapse of one species may be compensated for by increasing numbers of some other organism in the ecosystem. The disappearance of any species makes the web of life in the oceans weaker and less able to recover from shocks like global climate change, pollution, or overexploitation. Loss of marine biodiversity is impairing the ocean's capacity to provide food, maintain water quality, and recover from perturbations.

* * *

Many factors are responsible for the predicted collapse of numerous species of fish in freshwater, seas, and oceans. They include the capacity, range, and accuracy of fishing fleets, bottom trawling, aquaculture—particularly of carnivorous fish—and feeding fish to farm animals. The depletion of world fish populations globally has not prevented about 14 million tons of edible wild fish, such as anchovies, sardines, mackerel, and herring, from being ground up as fishmeal and fed to pigs and chickens each year. This amount of fish is more than six times that eaten by the entire U.S. population.

The productivity of marine areas today is only 10 percent of the maximum in earlier times. More than a quarter of the edible species have collapsed, and almost all edible species are projected to collapse within the next few decades if exploitation continues at present rates. The impending shortage of food cannot be compensated for by farming aquatic life; even the existing rate of production is not likely to continue for long. Modern marine fishing and aquaculture are surrendering future prospects for short-term gains.

6 * RESOURCES

In the final analysis, all forms of life require energy for their growth and sustenance. While all human activities impose a burden on the presently available or stored energy, food—the primary requirement for our survival—consumes more energy than all other activities. This fact is not usually appreciated in the developed world because of the availability of foods of various kinds in abundance.

The ultimate source of all energy on earth is the radiation that is received from the sun, with a tiny contribution from nuclear energy. Modern lifestyles require an enormous amount of energy that cannot be found in presently available and harvested solar energy and have become heavily dependent on fossil fuels, solar energy that was converted into useful forms millions of years ago. In the United States, for example, it is estimated that 85 percent of the total energy used is derived from fossil fuels.

In addition to energy, the production of food has a number of ancillary requirements. These include suitable environmental conditions (weather and temperature patterns), suitable land for farming, topsoil rich in organic matter, a regular supply of water, and the cooperation of insects and microscopic organisms. The environment is crucial; most crops are productive only within a narrow range of temperatures. Even a small fluctuation will adversely affect their output, and large variations from the norm will simply kill them. Land suitable for farming must have soil that allows penetration by the roots of plants and that contains minerals essential for their growth.

While solar energy is essential for the photosynthesis that is the very basis of all forms of plant life, agriculture requires inputs of

various other forms of energy. Work by humans, animals, or machines is required for tilling the soil and harvesting the crops. Fertilizers, insecticides, and other chemicals that are routinely applied to the crops are petrochemical products that also depend on these stored sources of energy.

The choice of foods we eat determines the amount of energy and resources that are used on a continuing basis. The amount of food available these days is not sufficient to feed the human population of about seven billion. This shortage of food will become more acute because the population is increasing and food preferences are changing in favor of energy-rich animal products. Hence, it is important to compare the resources needed to produce various animal products with the resources needed to produce plant foods for direct human consumption. By eating plant products, we are using the primary product of the energy harvested by plants. Feeding these items to animals and then consuming animal parts is a secondary process, with a concomitant loss of energy. A greater dependence on foods of animal origin increases the burden on planetary resources in two ways: the energy required to produce these foods is greater, and they consume greater amounts of water and other resources than plants do.

EFFICIENCIES OF PRODUCTION OF ANIMAL-BASED FOODS

All forms of life represent stored energy and have built-in mechanisms for converting energy from one form to another. For example, insects may use the energy content of leaves to build their bodies and to fly, and the insects then are eaten by higher-order animals such as birds that depend on this nourishment for their sustenance. While part of the food ingested by any organism is used to build its body, particularly in the early part of its life, a major portion at all ages is required for basic necessities—maintenance of the body, survival, defense, locomotion, and reproduction. All higher organisms and many lower organisms have needs related to socializing, communication, and other activities, all of which require expenditure of energy. Conversion of energy from one form of life to another necessarily results in a substantial loss at each step. In essence, farm animals consume a large amount of plant products that have low energy density and, after using a substantial portion of the energy and proteins for their own activities and sustenance, provide the rest to us as food. Even though the total protein and energy content in the animal-based foods is much less than that ingested by the animals, the animal products have a greater concentration of these items, which are the main reasons for their appeal.

Some of the energy consumed by animals goes into building their skeletons and body parts such as skin and hair. Since humans do not consume these animal parts, they represent a waste of energy in terms of available food for humans. Even in genetically modified species that are designed to be fatter at the expense of being healthier, the meat obtained after slaughter rarely exceeds 40 percent of the animal's weight. The ratio of energy in the ingested food to the energy present in the body of the animal depends on the species, the breed, the type of feed it consumes, and the environment in which it is kept. There are large variations among breeds—some breeds of steer have more flesh in their bodies; different breeds of cows do not give the same amount of milk under identical conditions; and some breeds of hens lay more eggs than others. These variations affect the output produced by different animals from the same amount of feed.

Food-conversion Efficiencies of Livestock

For any animal, the efficiency with which feed gets converted to useable food for humans depends on many factors, including type of feed and living conditions. As noted earlier, hormones and drugs are often used to increase the amount of food produced by animals. Modern industrial facilities use substantial proportions of grains in the feed of livestock to speed body growth. While this diet increases the energy in the input required to produce meats, a rich diet and administration of chemicals decrease the time needed for an animal to reach maturity: beef cows in feedlots put on enough weight to be ready for slaughter when they are younger than two years, whereas a grass-fed steer ingests feed with lower energy density and takes almost twice that time to mature. Confining animals in feedlots and breeding crates saves space, decreases the cost of maintaining a suitable environment, and prevents the animals from spending energy on muscular activities.

Because there are large differences among intrinsic energy-conversion efficiencies of various breeds of animals within a species as well as variations in their living conditions and the feed given to them, it is not possible to assign a single number to the efficiency with which any species of animal converts its feed into food for humans. However, some general statements can be made about the relative efficiencies of different farm animals. Milk is produced from feed in a rather straightforward process facilitated by the microbial population in the forestomach of cows. The conversion takes a very short time, in no case more than a day or two; the amount and type of food given to a dairy cow will affect the production of milk within a very short time. A dairy cow has a productive life of about five years, and thus the cost of maintaining the replacement herd is not very high. The higher efficiency factors in milk production apply to eggs as well. Hens lay eggs almost on a daily basis, roughly on nine days out of ten; their feed is quickly converted to eggs. A hen may continue producing eggs for two years, thereby reducing the expense of keeping the replacement stock.

The production of meats is different. Converting grains, hay, and other materials to the flesh of animals is a complicated and protracted process that involves multiple steps. In its lifetime, the animal has to take care of its bodily requirements, including those that are not associated with growth, such as development of reproductive abilities. Since an animal has to be killed to get access to its flesh, a replacement herd has to be maintained at all times, ready to take the place of those that have been slaughtered. Animals that

Table 6.1. Efficiencies of the production of foods of animal origin

	Milk	*Eggs*	*Chicken*	*Pork*	*Beef*
Food energy content (kcal/kg)	650	1,600	1,800	3,100	3,000
Protein content (%)	3.5	13	20	14	15
Energy conversion efficiency (%)	20-25	15-20	10-15	15-20	4-5
Protein conversion efficiency (%)	30-40	30-40	20-30	10-15	5-8

Source: Vaclav Smil, *Feeding the World: A Challenge for the Twenty-first Century* (Cambridge, MA: MIT Press), 2000, with permission.

are inherently more active have a faster metabolism that requires a greater expenditure of energy. While comparing different animals, it is useful to consider the lifecycle of the animals—that is, the size of the litter and how much time they take to reach maturity and reproduce. A steer requires more energy for daily activities than a pig and takes longer to mature; a cow gives birth to only one calf after a long gestation period, in contrast to a sow, which may give birth to as many as a dozen piglets after a shorter gestation period.

The efficiency of the production of animal-based foods may be defined as the ratio of food obtained from the animal to the feed given to it. There is no single yardstick to evaluate the contents of various foods; however, two aspects of great importance are their protein and energy content. Using these criteria, one may evaluate the protein conversion efficiency as the ratio of the protein in the final product to the protein in the feed and the energy conversion efficiency as the ratio of the total energy obtained in the animal-based food to the total energy content of the feed.

Both these numbers have large variations for animals raised in various settings and in different parts of the world. A number of authors have estimated such efficiencies for the industrial mode of production in Western countries for which data are easily available and where there are fewer variations in the methods of raising livestock. In general, it has been found that the production of milk and eggs is most efficient, while beef produces the lowest return for the amount of ingested feed. Although the production of all types of meat is very inefficient, the production of chicken meat is more efficient than the production of other meat. The protein conversion efficiencies for various animal products are estimated to be:

milk and eggs	35 percent
chicken	25 percent
pork	12 percent
beef	6 percent

The energy conversion efficiencies are:

milk	22 percent
eggs	17 percent
chicken	12 percent
pork	18 percent
beef	5 percent[1]

The energy conversion efficiency of pork is slightly greater than its protein conversion efficiency because pork contains a large amount of fat (lard).

Humans can directly subsist on many of the items fed to farm animals. Therefore, any efficiency less than 100 percent represents a loss from the perspective of the ecosphere's capacity to support the human population. While cattle can survive on hay, forage, seeds from which oils have been extracted, and agricultural refuse, the usual practice of the livestock industry is to feed animals grains and other high-value products that could be directly consumed by humans. In contrast to cattle, the feed of swine and chickens invariably consists of grains, so it overlaps to a greater extent with foods eaten by humans. According to a recent estimate, the cereals and pulses fed to livestock contain enough energy to feed more than three billion people on a purely vegetarian diet.[2]

OTHER REQUIREMENTS OF LIVESTOCK OPERATIONS

Although calculations of efficiency based on the feed ingested by animals and the product received for eventual human consumption is important, it does not capture the full ecological burden of producing animal-based foods. A comprehensive analysis must include other resources that are essential for the livestock industry to produce food, whether directly or indirectly, including fossil fuels, water, and land.

Fossil Fuels

CAFOs use energy from fossil fuels to maintain buildings and pumps that move manure, feed, and other items. Growing the modern varieties of corn, wheat, and rice that brought in the Green Revolution requires regular applications of fertilizers, insecticides, and herbicides, all of which are petrochemical products. Energy is used to transport animals and their feed back and forth over long distances and to operate slaughterhouses. Modern slaughterhouses work as assembly lines designed to process a very large number of animals that arrive from distant feedlots. Meat has to be kept frozen or refrigerated until it is used by consumers, and thus its storage requires more energy than storing grains. Because of all this expenditure of energy in various phases of livestock operations, calculating a simple ratio of the energy content of feed to that in the meat results in an overestimation of the efficiencies of producing various types of meat.

If we consider all the energy used to produce foods of animal origin, we find that producing meats and animal products uses anywhere from 2.5 to 50 times more fossil fuels than producing vegetable proteins[3] (the large range between 2.5 and 50 is due to the wide variety of methods of raising livestock and bringing their products to market). If all types of animal products are lumped together, the average fossil energy required to produce one kcal of protein is 25 kcal, more than 11 times greater than the energy required for the production of plant protein, 2.2 kcal.[4]

Although the total underground storage of fossil fuels is very large, it is not infinite, and extracting it will become increasingly difficult as its stocks are depleted. The use of fossil fuels also leads to the emission of numerous gases in various stages of the operation, including gases that cause pollution and carbon dioxide that contributes to global warming. The climate change accompanying this effect will have a direct bearing on the availability of food for human consumption.

Land

In the industrial mode of production, the space directly used by farm animals is very small, but a much larger area is required to grow feed for them and to provide all the other requirements for these operations. Because of their low conversion efficiencies, cattle need more feed per pound of body weight than other animals. However, at least some portion of their feed is obtained as byproducts when oil is extracted from seeds and when husks are obtained in the process of milling grains. The land requirement of these items should not be included in raising cattle, because these operations are independent of the livestock industry. Ruminants that are fed hay and forage will have different land requirements than those that have grains as their main feed component. Due to these variations, the land requirements of cattle in various industrial settings may differ by a factor of two or three. Vaclav Smil of the University of Manitoba has calculated the land area required to produce various types of animal-based foods after taking all factors into consideration. Milk scores highest here, just as it does in considerations of energy efficiency. It takes 1 square meter to produce milk that contains 1000 kcal of energy; eggs require 50 percent more land, and pigs slightly more than that. Production of chicken meat takes twice as much land as milk, and beef requires six times the area.[5] On the average, land requirements are roughly a factor of 6 to 17 greater for obtaining proteins from meat than for obtaining proteins from grains.[6]

Labor

A complete evaluation of resources needed to produce foods of animal origin has to include the amount of required labor, particularly because of large differences between growing grains and producing animal products. Among foods of animal origin, labor requirement is least for poultry production, because one person-hour of labor produces 25 kilograms of chicken meat. The main reason for this is that a factory operation can produce tens or hundreds of thousands of chickens using semiautomatic operations. Milk production is very labor intensive because of continuous involvement of workers, both in dealing with cows and handling milk through various phases, and produces only 2 kilograms for each person-hour of labor. The amount of labor required for the production of corn and other grains is much smaller, because crops require a lot of attention only during the planting and harvesting phases. On average, producing animal protein requires about 16 times more labor than producing proteins from plants.[7]

Water

While the need for fossil fuels will not limit the quantity of food produced by farmlands in the near future, the situation with water is completely different. Water scarcity is already limiting the production of food in some parts of the world, and the situation is getting progressively worse in many regions. In the semiarid climate of California, irrigated pastures used for growing grass for cattle constitute the single biggest usage of water in that area. When all other inputs are the same, the amount of available water determines the productivity of agricultural lands.

The amount of water directly consumed by livestock is much smaller than that needed to produce the feed. It takes at least 50 gallons of water to produce 1 pound of corn;[8] requirements for other grains and agricultural products are roughly of the same order. Just as for fossil fuels and other inputs such as land and labor that go into the production of animal-based foods, there are large variations in the water requirements in different types of operations due to various methods of housing, feed, climatic conditions, and ancillary operations. In general, it takes at least 10 times as much water to produce a pound of beef as a pound of grain, and in some situations, the ratio may be as much as 100 to 1.[9] Among meats, chicken requires the minimum amount of water, less than half that of beef.

Table 6.2. Estimated input of water required to produce 1 kg of forage crop or food

Crop/Food	*Liters/kg*
Potatoes	500
Wheat	900
Alfalfa	900
Sorghum	1,110
Corn	1,400
Rice	1,912
Soybeans	2,000
Broiler chicken	3,500
Beef	100,000

Source: David Pimentel et al., "Water Resources: Agriculture, the Environment, and Society," *Bioscience* 47(1997): 100, with permission.

Table 6.3. Requirements for producing foods of animal origin in the United States

	Requirement to produce 1 kg		
Animal product (1 kg)	*Grain (kg)*	*Forage (kg)*	*Fossil energy input/kcal output*
Lamb	21	30	57:1
Beef	13	30	40:1
Eggs	11	—	39:1
Pork	5.9	—	14:1
Milk	0.7	1	14:1
Turkey	3.8	—	10:1
Broiler chicken	2.3	—	4:1

Source: David Pimentel and Marcia Pimentel, "Sustainability of Meat-based and Plant-based Diets and the Environment," *American Journal of Clinical Nutrition,* 2003, 78(3), 660S–663S, with permission.

The amount of water needed to produce a pound of beef is frequently discussed in comparisons of vegetarian and meat-based diets. Since exact calculations of water consumption in all phases of operations are nearly impossible, one has to use estimates that may be inadvertently slanted to meet expectations. The beef industry maintains that 441 gallons of water are required to increase the weight of the cattle by one pound and 840 gallons of water are needed to produce a pound of beef,[10] since the carcass contains parts that are not useable as food. However, this estimate uses liberal deductions for water that evaporates during the process or is returned to the water table from agricultural lands and excrement of cattle. At the other extreme, David Pimentel at Cornell University estimates that more than 5,000 gallons of water are required to produce each pound of beef. His calculations assume that a beef animal consumes 100 pounds of hay and 4 pounds of grain per pound of beef produced and that these products are obtained from intensive agriculture.[11]

Arjen Hoekstra of the University of Twente and Ashok Chapagain of the World Wide Fund for Nature determined the amount of water required to produce various food items, including grains and animal products.[12] The virtual water content of products varies greatly from place to place, depending on the local climate, the technology used for farming, and the other steps involved in converting agricultural products to food items. Taking these factors into consideration, they obtained a global average for the amount of water required for each item. The production of beef consumes the most water, about 2,050 gallons per pound. Other types of meats require roughly 500 to 800 gallons of water per pound. Corn and wheat require 120 and 170 gallons per pound, respectively; the efficiency of water use was greater in the United States than in India, Australia, and Russia by up to a factor of three. It is now generally accepted that it takes at least 2,000 gallons of water to produce a pound of beef.[13] It is reasonable to estimate that an omnivorous diet requires five to ten times more water than a diet that is based purely on plant products.

HARVESTING MARINE LIFE AND AQUACULTURE

Fish use their feed with greater efficiency than livestock do. First, their bodies remain at the temperature of the water, so they don't spend energy to maintain a temperature different from that of their surroundings. Second, their bodies are streamlined and buoyant, so floating and moving require minimum effort; thus, fish do not have any large muscles that would consume a lot of energy.

Fish need diets composed of 35 to 40 percent protein—considerably more than poultry or mammals, which consume more carbohydrates than fish do. In contrast to agriculture, fishing does not require fertilizers or pesticides. The energy expended in catching wild fish greatly depends on the method of fishing. Coastal fishing by methods that do not require an excessive amount of fossil fuels gives a very high return per unit of energy used. Since the stock of coastal fisheries has been seriously reduced, a substantial portion of fishing is done these days with large factory-type trawlers that require fossil fuels for navigation and running the equipment. These trawlers usually travel over long distances to get the desired quantity of fish. As an extra burden on the fuel, the fish have to be kept frozen during the long trip back to the shore. To avoid having a large freezing facility, some vessels have canning equipment on board.

Due to increased solubility, the oxygen content of water increases with decreasing temperature. Other things being the same, the amount of dissolved oxygen determines the amount of aquatic life in any body of water. For this reason, the water around Alaska is much more productive than that of warmer regions. Out of the total fishery production of 4.4 million tons in the United States, more than half is obtained from Alaska. The average for all fish harvested in the United States is 27 kcal of fossil energy per kcal of fish protein produced; thus the input of fossil fuel for catching wild fish may be up to 14 times greater than the input needed for the production of vegetable proteins. If the energy expenditure in processing, a necessary part of trawler fishing, is included, the difference between the energy expenditures in fishing and growing vegetable proteins becomes even more pronounced. The amount of fossil fuel used in harvesting each kilogram of edible fish is so large that trawler fishing is not energetically or economically viable and depends on subsidies from governments for its very existence.

Aquaculture facilities have somewhat different resource requirements. Trawler fishing plays an important role, because it provides the fishmeal and fish oil required for raising carnivorous fish. Aquaculture facilities require substantial amounts of water that is returned to the environment in a highly polluted form. In ponds with intensive production, about a third of the water has to be changed daily; about half of it is freshwater, needed to maintain the optimum salinity.

DIET, MALNUTRITION, AND HUNGER

The human population, which is increasing by 77 million per year and approaching 7 billion, stretches the capacity of the planet to provide for our needs. The shift in food preference from plant foods to animal products that is taking place in most regions of the world greatly increases the rate of consumption of precious resources. This trend is likely to create a gap between the capacity of the land and the seas to produce food and the amount needed by the human population for its sustenance.

Food Shortages and Hunger

The simplest indicator of a discrepancy between the demand for and the availability of food is indicated by its price, which has been steadily increasing. According to data compiled by the U.N. Food and Agriculture Organization (FAO), the price index of food items is increasing rapidly: in 2000, the price indexes of cereals, oils, and sugar were 87, 72, and 105, respectively. The corresponding numbers in April 2008 were 284, 276, and 161, representing a very large increase in the price of basic necessities.[14] As shown in a report by the United Nations Environment Program issued in March 2009, the increase in food prices during the last few years has been greater than at any other time during the last century and has affected all food items. As a result of the increases, an additional 110 million people have been driven into poverty and another 44 million added to the undernourished population.[15] The index of food prices compiled by the FAO showed an increase of 9 percent in 2006 and 23 percent in 2007 and a surge of 54 percent in 2008,[16] triggering riots in many countries, including Egypt, Haiti, Cameroon, and Bangladesh. While fuel prices, which also surged in this period, subsequently declined slowly, food prices remained at a high level. This is not just a short-term price blip but the beginning of a major structural change in the world's food market.[17] The worldwide global recession beginning in 2008 did not cause a decrease in the price of rice, a major grain consumed by much of the world, and the tenacity of the price is an indication of a systemic shortage of food.

High food prices dismay even relatively well-off people, but they are truly devastating for poor people and for most people in poor countries, where purchasing food often takes a large fraction of a family's income. While a household in the United States spends less than 15 percent of its budget on food, families in most underdeveloped countries spend more than half of their disposable income to provide for this basic necessity, and this proportion may approach 70 or 80 percent.[18] When the proportion of income spent on food is so high, any increase in the price of food leads to malnutrition or even starvation. World cereal grain stocks are at an all-time low, food aid programs have run out of money, and millions face chronic shortages of food.

Although the incidence of hunger due to lack of resources is much lower in the United States than elsewhere, it is still significant and will increase as prices of food items increase. According to data compiled by the USDA for the year 2007, 7 percent of U.S. households (5.5 million adults and 8.8 million children) had low food security, which means they had to use various coping strategies such as federal assistance programs and food pantries to get the food they required. An additional 4.1 percent of households (8.2 million adults and 3.7 million children) did not have enough money or other resources to buy food all through the year.[19] Official figures released by the USDA in November 2009 indicate that there was a sharp increase in these numbers in 2008, when 8.9 percent of households (10.4 million people) were living with low food security and 5.7 percent of households (6.7 million people) were living with hunger and very low food security. Additional increases in the price of food items will drive more and more people into these two groups.

Resources Used to Produce Animal-based Foods

As we have seen, a diet with foods of animal origin as its main component makes a much greater demand on planetary resources than a diet of plant products. Compared with the production of proteins from plants, proteins from meat are obtained at the expenditure of:

4.4 to 26 times the amount of water
6 to 20 times the amount of fossil fuels
7 times the amount of phosphate rock
6 to 17 times the amount of land[20]

The ranges given for each of these resources are due to variations in local conditions and details of farming and livestock operations; the higher numbers generally refer to livestock production in CAFO-type facilities. Many of these resources are already in short supply, and their stock is continuously dwindling.

Keeping in mind that world cereal stocks are at an all-time low, it is useful to consider what proportion of grains and other agricultural products are used by the livestock industry. On a worldwide basis, more than one-third of the cereal production is fed to livestock; the total of all cereals, oilseeds, roots, and tubers that support farm animals is 744 million tons. In addition to cereals and grains, 28 million tons of seafood, about 30 percent of the total harvest of marine life, is included in the feed despite the fact that fisheries all over the world are being depleted at a rapid rate, threatening the lives and livelihoods of millions of people. Thus, a large proportion of the world's agricultural harvests end up in the digestive tracts of livestock, where the conversion of vegetable to animal proteins incurs losses of between 60 and 90 percent of the energy. The difference between the energy fed to the farm animals and that obtained from animal products, which represents the energy loss due to meat-based diets all over the world, is sufficient to feed 3.5 billion people on a purely vegetarian diet.[21]

The proportion of grains and other agricultural products fed to livestock is much greater in developed countries than in developing ones. In the United States, more than nine billion livestock raised each year to supply animal-based food are fed more than seven times as much grain as the American population consumes.[22] Food production systems use about 50 percent of the total land area, 80 percent of the freshwater, and 17 percent of the fossil energy in the United States.[23]

Among the grains that are fed to farm animals, corn tops the list. According to the Corn Growers Association, 60 percent of the corn harvested in the United States is fed to farm animals, but other estimates claim the percentage to be even greater. Soybeans are the next most common grain fed to livestock; the proportion of the U.S. harvest of soybeans that goes into feed is estimated to be as high as 70 percent. The amount of grain fed to livestock in the United States is sufficient to feed about 840 million people following a plant-based diet.[24] This number would be even greater if pulses, legumes, and tubers fed to farm animals are included in the calculation. Furthermore, if the entire world paralleled the energy intensiveness of the U.S. agricultural production system, proven worldwide oil reserves would be exhausted in 12 years just to provide food for the population.[25] The food habits of other developed countries are somewhat similar and represent a comparable loss of energy in the conversion of plant to animal products.

TRENDS IN FOOD REQUIREMENTS

Recent years have seen a substantial increase in the population of the middle class in developing countries. Increasing affluence is almost always accompanied by a demand for more meat, poultry, seafood, and dairy products. A variety of factors contribute to this change. Animal products are rich in fats and proteins and give a greater feeling of satiety. They also provide all the ingredients required for the sustenance of the human body in one package and are often considered to be healthier than plant-based foods. In today's globalized world, people in developing countries often try to emulate the lifestyle of those in developed countries. During the last half-century, the Japanese diet has incorporated many Western fast food items, with meat replacing seafood. The same thing is happening in China today, the only difference being that Chinese people consume much more pork than other meats.

An increasing number of people in emerging economies are, for the first time, rich enough to start eating like Westerners. Since it takes between 700 and 1,000 calories worth of animal feed to produce a 100-calorie piece of beef, this change in diet increases the overall demand for grains and other plant products. The amount of meat eaten by an average Chinese person almost doubled during the 11-year period from 1991 to 2002. Although at 115 pounds, it is still substantially less than the 275 pounds consumed by an average person in the United States, it is increasing at a very rapid rate.[26] As a result of dietary changes, China has become a major importer of foods of all kinds. A net soybean exporter until 1995, China has since increased its import every year, reaching 30 million tons in 2007.[27] China also established import targets of 7.2 million tons of corn, 9.6 million tons of wheat, and 5.3 million tons of rice. Since all countries compete for grains in the international market, even a small shift in China's demands, caused by drought or changing food preferences within the country, can have a devastating effect on the global price and supply of food.

The per capita consumption of meat is projected to increase from 82 pounds per year in 2000 to over 115 pounds by 2050.[28] Due to the increase in population during this period, this change will require that more than half of the cereals produced worldwide be used by the livestock industry. Even if the adverse effects of land degradation and climate change are not taken into account, growing this amount of cereal will require an additional 50 million acres (120 million hectares) of cropland by 2030, an area twice the size of France and one third that of India. Almost all the land that can support agriculture is already being used for farming; the unused terrain is generally not suitable for growing crops on a regular basis. In addition, precious farmlands are being lost to urbanization. In Asia, nearly 95 percent of the potential cropland has already been utilized. Increases will be difficult to manifest in large parts of sub-Saharan Africa due to political, socioeconomic, and environmental constraints. Increases in cropland may be possible to achieve in Latin America by cutting down rainforests; to satisfy the demand for meat, rainforests in Brazil, Bolivia, and Paraguay are being cut down for conversion into pastures and croplands. Clearing the rainforests will accelerate climate change and loss of biodiversity that may adversely affect crop yields in the future.

DIMINISHING RESOURCES

At the same time that the demand for food is increasing rapidly, there are indications that the growth in the supply of food will slow down or even be reversed. Crop yields in developing countries have fallen dramatically for at least a decade, with diminishing returns for various agricultural inputs. Climate change has disrupted crop growth and water distribution. Up to 25 percent of the world's food production may become "lost" during this century as a result of climate change, water scarcity, invasive pests, and land degradation. Subsidies given to farmers in rich countries cause global trade distortions when the produce of rich countries is dumped into the markets of poorer ones, wrecking the livelihoods of small-scale farmers. This imbalance also leads to decreased local food production and exposes the world to the inherent dangers of monoculture.

Biofuels, once perceived as the green alternative to fuel, have recently been discredited. If all aspects of the production of biofuels are taken into account, the adverse environmental effects of this source of energy are much greater than those from fossil fuels. Biofuels also decrease the amount of food available for consumption by humans. Americans now burn enough corn in their gas tanks to cover the import needs of 83 food-deficit countries.

Serious strains are evident on many precious resources, including land, water, and phosphorus, that are needed to provide food to the human population. Per capita arable land is decreasing everywhere due to increasing population and urbanization; worldwide, there is now estimated to be half an acre per person, with a much smaller area in the most populous countries of the world. A shortage of water is already limiting the harvest of crops in some of the most productive regions of the world. Phosphorus is an essential component of chemical fertilizers and is also required in the feed of livestock. It is primarily obtained from phosphate rocks found in very few places, including Morocco, Christmas Island, and the island of Nauru in the Pacific Ocean. There is no substitute for the critical role of phosphate in plant development and production and, since its stocks are found in few locations and are very limited, the exponential growth in global food production could lead to an acute shortage of phosphate within two or three decades. This projected shortage is primarily occurring because, unlike in previous eras, phosphates from human and animal waste are not recycled properly. As a result, they end up in bodies of water, eventually making their way to the oceans.

* * *

It appears that the era of cheap food, brought about by the Green Revolution in the last three decades of the twentieth century, is over. The population of the world increased from 4 billion in 1974 to 6.8 billion in 2009, while simultaneously intensive agriculture using fertilizers, pesticides, and herbicides degraded farmlands and reduced their productivity. Food, the fundamental determinant of health, is unaffordable to an increasing proportion of the world's population. The world is expected to have at least two billion more people to feed by 2050. Emerging economies are not only eating more but also eating more meat. Diverting grains to feed livestock is continuously decreasing the supply of food for humans and also consuming precious resources at a rapid rate. According to experts, the current shortages and price hikes are not a phenomenon that will end in a few years but one that will continue for the foreseeable future. Widespread malnutrition and starvation will result.

The present global trends in energy consumption are patently unsustainable—environmentally, economically, and socially. The future of human prosperity depends on reducing the consumption of energy and the environmental damage caused by present lifestyles and finding environmentally benign methods of obtaining the desired items from the ecosystem. The selection of the food we eat is a serious choice that determines not only the sustainability —that is, the capacity of the planet to continue to provide for the needs of future generations—but also the availability of food to people with limited resources living right now. A dispassionate analysis of the present situation shows that dramatically reducing meat consumption in developed countries and reversing the recent trend toward the consumption of meat in developing countries appears to be the only solution to stave off the disastrous consequences of our dietary choices. A major shift toward vegetarianism by a large segment of the population will reverse land degradation, limit the loss of biodiversity and resources, and protect the food production platform of the planet for future generations.

7 * HEALTH

Scholars in the field of diet and nutrition define two kinds of diets in relation to human health. An "adequate diet" provides sufficient nutrients and energy for human maintenance, growth, and reproduction. An "optimal diet," on the other hand, in addition to providing the required nutrients, also promotes health and longevity by reducing the risk of chronic diseases.[1] The typical Western diet, based on meat, potatoes, and bread made from refined flour, amply fulfills the criteria for an adequate diet. Since the average person living in a developed country consumes about 3700 calories each day—far greater than the 2500 calories suggested by the nutritional guidelines—the energy requirement is more than satisfied. However, epidemiologists have estimated that if those same persons consumed an optimal diet rather than an adequate diet, many of them could avoid the chronic diseases that cause mortality and morbidity among millions of people or could at the least delay the onset of disease. An optimal diet has fewer calories and is rich in various plant products.

CHRONIC DISEASES

Chronic diseases are, to a major extent, lifestyle diseases that develop over a period of years, slowly degenerating or attacking some parts of the body. An understanding of the role that diet plays in the progression of these diseases, or in their prevention, requires an understanding of human physiological events that take place during the early stages of these afflictions. These noncommunicable (noncontagious) diseases take a heavy toll on people's lives. They are:

cardiovascular disease
cancer
type 2 diabetes mellitus
neurodegenerative conditions such as Alzheimer disease and Parkinson disease

Chronic diseases are the leading causes of premature death and disability in the Western world. They account for 70 percent of all deaths in the United States, about 1.7 million per year, and also cause major hardship in the daily lives of nearly 1 in 10 Americans, about 25 million people.[2] Chronic diseases adversely affect people's quality of life; for many people, they cause total or partial disability for many years. These afflictions also place an enormous strain on family and community finances. Local and national economies suffer from the labor lost due to death and illness, the high direct medical costs, and the investment of time by caregivers. Data kept by the U.N. World Health Organization indicate that chronic diseases gave rise to approximately 60 percent of the 56.6 million deaths reported around the world in 2005. It has been projected that by the year 2020, chronic diseases will account for about 75 percent of deaths worldwide. The incidence of chronic diseases in developing countries, which had been much lower than in developed countries, is rapidly rising and in middle-income countries has started approaching the rates in the developed world. Eighty percent of Chinese people die as a result of noncommunicable diseases, often tied to diets high in animal products and sugars.[3]

The complexity of the human body and differences in the environmental conditions that each person encounters in daily life make it difficult to pinpoint the exact cause of most chronic diseases; however, enough information has been obtained about the sequence of events that lead to the diseased state to understand the role of some foods in accelerating the progression of chronic diseases and that of other foods in slowing down the process. Important epidemiological information can be obtained by monitoring the frequency of chronic diseases in people who move from regions of the world with a low incidence of chronic diseases to the developed world, where they adopt the food habits and lifestyle prevalent in the host country. It has been found that such groups suffer from degenerative diseases at almost the same rate as the native population. Observations of this type indicate that the contribution of genetic factors to the development of various chronic diseases is rather small; most of the diseases are caused by lifestyle and eating habits.

Establishing a cause-and-effect relationship between the intake of a particular food and a specific disease is nearly impossible, because it would require regimentation of the diets of genetically identical people over a period of many years, during which only one factor is changed while all others are kept the same, for the duration of the investigation. To overcome these constraints, studies to determine the relationship between diet and disease are done by collecting statistical data for large groups of people.

Due to large genetic and lifestyle differences within any selected group of people, and because, given freedom of choice, people seldom adhere to a specified diet over a period of many years, the accuracy of conclusions increases with the number of participants. Diseases such as cancer and cardiovascular disease (CVD) have long latency periods before the appearance of clinical symptoms, and genetic variations make some people susceptible and others resistant. In addition, the onset of chronic diseases in any group of persons during a period of a year or two is rather small. For all of these reasons, meaningful conclusions (called statistically significant results) can only be drawn from studies that involve thousands of persons whose habits and incidence of disease are followed for many years.

A number of cohort studies have helped define the association between degenerative diseases and lifestyle factors and have provided useful information about the effects—both harmful and advantageous—of various foods on the development of chronic diseases. Two prominent studies of this type that have been running for many years are the European Prospective Investigation in Cancer and Nutrition (EPIC) and the Nurses' Health Study in the United States. Under the umbrella of EPIC, investigators at many European institutions have been gathering information since 1992 about the effects of specific dietary components on long-term health. These studies involve over half a million people in ten European countries.[4] In the United States, the Nurses' Health Study was started in 1976 and has conducted investigations into lifestyle determinants of CVD and cancer in 121,700 registered nurses living in 11 states.[5] In addition, a study of 97,000 Seventh-day Adventists in the United States and Canada has been examining the effects of a vegetarian diet on the onset of chronic diseases.[6] A few other large-scale studies are following dietary patterns and the frequency of chronic diseases in other parts of the world.

Cardiovascular Disease

CVD can lead to heart attack and stroke which, together, lead to the deaths of more people than any other disease. In the United States, about 29 percent of deaths are caused by heart disease each year. Epidemiological studies have established a link between the level of cholesterol in blood and the likelihood of suffering from CVD. The blood contains aggregates known as LDL (low-density lipoprotein) cholesterol and HDL (high-density lipoprotein) cholesterol, which are synthesized in the liver. Both types of particles perform essential functions in the body and are made of fats, proteins, and cholesterol. LDL particles transport these components to muscle, heart, and other body tissues, where they are required by dividing cells. HDL particles take cholesterol and fat to the liver for recycling and excretion. Blood also contains a class of compounds known as free radicals, which are highly reactive and can easily oxidize other constituents of blood. These particles are produced by the immune system and various processes in the body to kill invading pathogens and to mark damaged tissues for removal from the body. However, free radicals have the potential to damage important parts of cells, including proteins, membranes, and DNA. Cellular damage caused by free radicals has been implicated in the development of many diseases, including CVD, cancers,[7] and neurodegenerative disorders.[8] Oxidative damage caused by free radicals may be prevented by antioxidants, which are present in body fluids and tissues.

Unlike HDL cholesterol, some components of LDL particles are oxidized by free radicals, and the oxidized cholesterol is deposited on the walls of blood vessels, a process known as atherosclerosis. The tendency of LDL cholesterol to become oxidized by free radicals makes it the "bad cholesterol," although it normally performs an essential function in the body. Oxidation of LDL is a major factor in the promotion of coronary heart disease.[9] The overgrowth of tissues caused by atherosclerosis constricts the blood vessels and reduces the flow of blood, which may lead to chest pain (angina), particularly during exercise or hard work, when there is a greater need for blood.

Over time, an atherosclerotic plaque may rupture, causing thrombosis, in which a large blood clot forms in the blood vessel. If this clot stops the blood supply from reaching the heart, a heart attack results; if it stops the blood supply from reaching the brain, it causes a stroke. The amount of cholesterol in the blood is determined both by the genetic makeup of the individual and lifestyle factors such as diet and level of physical activity. Since cholesterol is synthesized in the body as well as taken in through food, dietary cholesterol is only one of the factors that determine the amount of cholesterol present in the blood.

Diabetes

A person develops type 2 diabetes mellitus when the beta cells in the pancreas cannot make enough insulin to meet the requirements of the body. Type 2 diabetes is often associated with being overweight or obese. Obesity is both a cause and an effect of diabetes. The pancreas may not be able to make enough insulin to meet the bodily needs of an overweight person, thus resulting in diabetes. At the same time, energy from food items cannot be absorbed by the cells of the body in the absence of a sufficient amount of insulin, and hence the starved cells keep giving signals to get more food, causing obesity.

Diabetes is a dangerous disease in itself, and it also increases the risk of heart disease. Most of the health problems that people with diabetes have are caused by elevated levels of sugar in the bloodstream that damage organs of the body, including the heart, lungs, liver, and eyes. The number of persons suffering from diabetes is estimated to be about 200 million worldwide. The prevalence of diabetes is rapidly increasing in the United States: the number of persons afflicted with this disease more than tripled during the 28-year period from 1980 to 2008, increasing from 5.6 million to 18.1 million.[10] Another 5.7 million people probably have this disease without knowing it. According to projections by the Centers for Diseases Control and Prevention (CDC), as many as one-third of adult Americans could suffer from diabetes in 2050.

Cancer

Cancer is the uncontrolled growth and spread of abnormal cells in the body. Cancer begins with damage to the DNA of a cell; this damage may occur due to exposure to external chemicals such as cigarette smoke or pollution, ingestion of certain compounds in food and drink, or exposure to the harmful effects of nuclear radiation, x-rays, ultraviolet rays, or cosmic rays. The rate at which such damages to DNA occur in everyday life is very large; however,

Table 7.1. Cancer incidence in North America, per 100,000 people, all ages, 2007

Male		*Female*	
Prostate	156.9	Breast	120.4
Lung	80.5	Lung	54.5
Colon and rectum	52.7	Colon and rectum	39.7
Urinary bladder	36.0	Uterus	24.1
Non-Hodgkin lymphoma	22.6	Non-Hodgkin lymphoma	15.7
Melanoma of skin	23.5	Melanoma of skin	15.4
Kidney and renal pelvis	20.8	Thyroid	17.2
Oral cavity and pharynx	16.1	Ovary	12.2
Leukemia	15.0	Kidney and renal pelvis	10.9
Pancreas	13.2	Pancreas	10.2

Source: U.S. Centers for Disease Control and Prevention, http://apps.nccd.cdc.gov/uscs/toptencancers.aspx.

our bodies have excellent built-in defenses that prevent most of this damage from proliferating. As a first line of defense, the new DNA created at the time of cell division is eliminated if it does not match the original blueprint. But the mutant DNA may survive this detection protocol and may proliferate, if the damage occurs while the DNA is replicating or during other vulnerable moments in a cell's life cycle. There are additional safeguards that eliminate the damaged cells that escape the initial detection process.

A tumor or undesirable growth occurs in rare cases when a mutant cell with damaged DNA is able to escape all the body's defenses. Benign tumors are characterized by slow, self-limiting growth, and they do not take over the whole site or spread into surrounding tissues. Many cancerous tumors, on the other hand, are characterized by uncontrolled growth. Cancer cells can also break away from a malignant tumor and enter the bloodstream or the lymphatic system and spread to other parts of the body, a process known as metastasis. The prevalence of cancer is indicated by the statistic that one-half of males and one-third of females living in the United States will have a diagnosis of cancer in their lifetime. About 1.3 million new cases of cancer are diagnosed each year, and it is the second leading cause of death in the United States.

Neurodegenerative Diseases

In neurodegenerative diseases, neurons in the brain that make it possible for people to control movements, process information, and make decisions stop communicating with each other, causing serious physical and mental impairment. More than five million people in the United States are currently living with Alzheimer disease, which involves loss of memory and inability to perform everyday functions. That number is projected to increase rapidly due to the aging population, with an enormous cost to the health care system and a great burden on the families of individuals suffering from this terrible disease. Parkinson disease impairs the affected person's motor skills, speech, and other functions. About half a million people suffer from Parkinson disease in the United States, and 50,000 new cases are reported annually. Free radicals present in the blood may play a significant role in the incidence of these diseases. The brain is highly susceptible to oxidative damage, because the plasma membranes of neural cells contain unsaturated fatty acids that can be easily oxidized by free radicals.

A different type of neurodegenerative disease is mad cow disease, in which the brain, spinal cord, and retina of cattle with the disease contain misfolded proteins called prions. It is believed that eating the meat of an animal that suffered from this disease may cause the human variant of mad cow disease, known as Creutzfeldt-Jakob disease (CJD). It progresses rapidly and leads to death within one year. Since 1990, the USDA has had an aggressive surveillance program to ensure detection and swift response to CJD in this country. To prevent the spread of this disease among cattle, the body parts of cows that may contain prions, such as the brain and spinal cord, are not fed to other cattle.

Obesity

An adult with a body mass index (BMI) between 25 and 30 kg/m^2 is classified as overweight, and an adult with a BMI greater than 30 is considered to be obese. While obesity is not classified as a chronic disease per se, it increases the likelihood of contracting chronic ailments and aggravates the physical impairments caused by them. The health consequences of this condition range from increased risk of premature death to serious chronic conditions that reduce the overall quality of life. Obesity, particularly abdominal obesity, is not only a major risk factor for type 2 diabetes but also for other noncommunicable diseases such as CVD and cancers of the breast, colon, prostate, kidney, and gallbladder. Chronic overweight and obesity contribute significantly to osteoarthritis, a major cause of disability in adults.

In an analysis reported by the World Health Organization in 2002, approximately 58 percent of diabetes, 21 percent of heart disease, and 8 to 42 percent of certain cancers globally were attributable to people having a BMI above 25. A study that followed more than 100,000 persons under the rubric of the Nurses' Health

Study and the Health Professionals' Follow-up Study for a period of 10 years found that the incidence of diabetes, gallstones, hypertension, heart disease, and colon cancer increased with the degree of excessive weight in both men and women. The risk of developing chronic disease was evident even among adults in the upper half of the healthy weight range (BMI of 22.0 to 24.9), suggesting that adults should try to maintain a BMI between 18.5 and 21.9 to minimize their risk of chronic disease.[11]

There are more than 1 billion overweight adults in the world, of whom 300 million are obese. Obesity has reached epidemic proportions globally and is a major contributor to the overall burden of chronic disease and disability. Its rates are increasing rapidly in all parts of the world. Rates have risen three-fold or more in some areas of North America, the United Kingdom, Eastern Europe, the Middle East, the Pacific Islands, Australia, and China during the last few decades.[12] Obesity in the United Kingdom increased by 30 percent in men, 40 percent in women, and 50 percent in children during a 10-year period.[13] According to the CDC, roughly two-thirds of the adult population in the United States is either overweight or obese. About 17 percent of children and teenagers between the ages of 6 and 19 are obese.

The rate at which obesity is increasing in the United States can be gauged from the fact that in 1990, no state in the country had a level of obesity equal to or greater than 15 percent. In 2009, only one state (Colorado) had a incidence of obesity less than 20 percent. Thirty-three states had a level equal to or greater than 25 percent, and eight of these states (Alabama, Arkansas, Kentucky, Louisiana, Missouri, Oklahoma, Tennessee, and West Virginia) had an incidence of obesity equal to or greater than 30 percent.[14] The percentage of the population with a BMI greater than 25 has steadily increased over the years in males and females of all ages, all racial and ethnic groups, and all educational levels. The number of overweight persons is increasing in developing countries even faster than in the developed world. In addition to increasing the incidence of chronic diseases and decreasing the life span, obesity also adversely affects a person's quality of life.

A MEAT-CONCENTRATED DIET AND HEALTH

An important reason for the increasing demand for meat and dairy products is the common perception that these foods are better for overall health than plant items, because they provide nutrients and energy in a concentrated form. A typical meat dish contains 10 to 25 percent protein (the exact amount depends on the type of meat and method of preparation). While the protein content of a meat-based diet is somewhat greater than that of a vegetarian diet based on grains and legumes (pulses, beans, and all nitrogen-fixing plants are legumes), the important difference is that the protein in meats is properly balanced—that is, its amino acid constituents are in roughly the same proportion as those found in human organs and hence are completely and easily assimilated into the human body. Obtaining all the amino acids in the desired proportion with a vegetarian diet may require eating food from different categories. Unless care is taken to mix vegetable proteins in appropriate proportions, it is necessary to divide the protein contents of vegetarian foods by 1.4 to account for their lower bioavailability.[15] The Institute of Medicine recommends that adults get 0.8 gram of protein

Table 7.2. Protein, fat, and cholesterol content of some common foods

Food item (100 g)	*Protein (g)*	*Total fat (g)*	*Saturated fat (g)*	*Cholesterol (mg)*
Hamburger	23.1	21.8	8.8	84
Hot dog	11.2	29.6	11.7	53
Pepperoni	22.7	44.0	14.9	105
Chicken breast	15.8	17.7	3.8	44
Chicken thigh	22.9	15.7	4.1	132
Pork chop	24.8	3.5	1.2	52
Pork sausage	19.4	28.4	9.1	84
Eggs, scrambled	13.1	5.6	1.1	65
Whole milk	3.2	3.2	1.8	10
Skim milk	3.4	0.2	0.1	2
Cheese	24.9	33.1	21.2	105
Whole wheat flour	13.7	1.9	0.3	0
Corn, dried	14.5	10.6	2.0	0
Soybeans	36.5	19.9	2.9	0

Source: U.S. Department of Agriculture, Agricultural Research Service, http://www.ars.usda.gov/Services/docs.htm?docid=6282.

for each kilogram of body weight, that is, 1.28 ounces of protein for each 100 pounds of body weight.[16] Surveys have shown that the average American eats almost twice the daily nutritional requirement of proteins and that 70 percent is from animal products.

All animal products contain saturated fats and cholesterol (described below). The amount of total fat in meat depends on the type of meat and its trimmings. Typical beef dishes contain 20 to 45 grams of fat, a substantial part of which is the saturated variety, per 100 grams of the food item. Although chicken has a somewhat smaller amount of fat, it often has more cholesterol than beef or pork. Sausage, in general, has more fat than other types of meat. Meat also contains iron and phosphorus in forms that are easily accessible to the body, in addition to some trace metals. Animals do not produce any of the micronutrients that are present in their meat; instead, they obtain these nutrients from their feed and metabolize them into a somewhat different form.

Possible Harmful Effects of the Consumption of Meat

Eating meat may have harmful effects on health for a variety of reasons. The consumption of meat increases the load of saturated fats and cholesterol in the body, with a consequential adverse effect on cardiovascular health. For reasons that are not properly understood, the intake of meat also produces harmful substances in the gastrointestinal system. Cooking meat, particularly at high temperatures, generates a few types of chemicals that are known carcinogens (cancer-causing agents). Finally, meat can easily harbor and promote the growth of bacteria and pathogens that may eventually infect the human body if they are not killed during the cooking process. The adverse effects of these items on human health depends on the type and amount of meat ingested by the individual.

A number of studies have indicated that the frequent consumption of meat increases the chances of contracting some chronic diseases. The largest study of this kind, conducted by scientists at the National Cancer Institute in the United States, involved more than half a million people between the ages of 50 and 71 whose eating habits and incidence of chronic diseases were followed for a period of 10 years. It was found that eating the equivalent of a quarter-pound hamburger daily (28 ounces a week) increased the risk of dying from cancer by 22 percent, while the risk of dying from heart disease was higher by 27 percent as compared with those who ate less than 5 ounces of hamburger each week.[17] Another large cohort study involving half a million men and women in the United States found a positive association between the risk of cancers of the colon, rectum, esophagus, and lung and the intake of red and processed meat.[18] It was estimated from these studies that approximately 35 percent, with a range from 10 percent to 70 percent, of cancers can be attributed to diet; thus, the contribution of meat to the incidence of cancer is comparable in magnitude to that of smoking, which is 30 percent with a range of 25 percent to 40 percent.

The relation between the consumption of meat and cancer has been established most unambiguously for the case of colorectal cancer (CRC), that is, cancers of the colon, rectum and appendix. CRC is the third most common cancer, with a million cases throughout the world. It has an incidence rate of 54 per 100,000 and a death rate of 22 per 100,000. In the United States, CRC is the second leading type of cancer and results in about 11 percent of all cancer-related deaths. In 2010, 152,500 new cases were diagnosed, and CRC was the cause of 55,000 deaths.[19] The five-year relative survival rate is 90 percent for people whose colorectal cancer is treated at an early stage; unfortunately, only 37 percent of cases are found early. The incidence of colorectal cancer is approximately tenfold higher in developed countries, where the intake of meats is greater than in developing countries. CRC was very rare in Japan in the 1960s, but there has been an almost fivefold increase in this disease in Japanese men during the last 30 years; its occurrence in Japanese men ages 55 to 60 is now greater than that of men in the United Kingdom.

A number of large-scale cohort studies that included thousands of participants have established that a daily intake of 3.5 ounces of red meat increases the risk of colorectal cancer by about 20 percent, while a daily intake of only 1 ounce of processed meat increases the risk of colorectal cancer by about 50 percent,[20] indicating that processed meat is much more potent in causing these diseases. In these studies lamb, beef, and pork were considered to be red meats, and sausages, burgers, ham, bacon, and cold cuts of all kinds were considered to be processed meats. A study that reached the same conclusion followed 478,040 men and women from 10 European countries for a mean follow-up period of 4.8 years.[21] It found that the contribution of diet to CRC could be as high as 80 percent.

Cooking meat at high temperatures, or even the natural process of digesting meats, may produce chemicals that are known carcinogens. Three such classes of chemicals that are not present in uncooked meats are heterocyclic amines (HCAs), N-nitroso compounds (NOCs), and polycyclic aromatic hydrocarbons. HCAs are produced when muscle meats such as beef, pork, poultry, or fish are cooked at high temperatures, or even at moderate temperatures, for a long time. These chemicals are formed when amino acids in proteins combine with creatinine, a chemical found in muscles, at elevated temperatures. NOCs are produced by the interaction of nitrites (which are often added to processed meats as a preservative and also to impart a reddish-pink color to red meats) with iron-containing proteins in red meat during normal digestive processes in the body. Iron in vegetarian foods and proteins from sources other than red meat do not produce NOCs.[22] The third class of potent carcinogens, polycyclic aromatic hydrocarbons, is produced when meat is cooked over a flame and its juices and fat stick to the meat, something that often occurs while grilling meats. These three classes of chemicals combine with DNA in the body to cause mutations that may lead to the growth of cancer if not eliminated by the defense mechanisms in the body.

The risk factors for diabetes and CVD—known as the metabolic syndromes—are high blood pressure, elevated triglycerides, high levels of total cholesterol with low levels of HDL cholesterol, and high fasting glucose levels. The importance of these metabolic syndromes is that they are often precursors of chronic diseases. While, as noted, variations in lifestyle and genetics and the time lag before clinical symptoms appear make it difficult to establish a direct correlation between the intake of meat and CVD, the effect of a diet rich in meat on these metabolic syndromes is direct and can be easily established. A number of studies that followed the dietary habits of thousands of persons have found that regular intake of red and processed meat increases the risk of developing metabolic syndromes by 25 percent.[23] A Coronary Artery Risk Development in Young Adults study involving 5,115 participants found a positive association between the consumption of red and processed meat and elevated blood pressure of participants. A similar relationship was found in a study of 41,541 predominantly white U.S. female nurses aged 38 to 63 years.[24] The consumption of plant foods—especially whole grains, fruits, and nuts—and milk and low-fat dairy products kept blood pressure within the normal range.[25]

Some of the damaging effects of a diet rich in animal products are caused by their saturated fat and cholesterol content. The fatty acid composition of different types of meats depends on a number of factors, including the species of the animal, the breed within the species, the type of feed on which it was raised, the cut of meat, and the method of cooking. In general, about half the fat in beef, lamb, and pork is saturated. Ingestion of saturated fats by humans increases the level of cholesterol in the blood, which is a risk factor for cardiovascular disease. Even though the correlation between the intake of animal products and CVD is somewhat weaker than for colorectal cancer, its effect on public health is much greater due to the large number of deaths and disabilities caused by heart attacks and strokes.[26]

Some studies have suggested a weak but statistically significant link between the intake of animal fats and hormone-related breast and prostate cancers. A study that followed 90,655 premenopausal women for eight years found that those who consumed the greatest amount of fats of animal origin, including dairy products, had greater incidences of breast cancer. The intake of vegetable fats was not correlated with any increase.[27] Prostate cancers are of two varieties. The slow-growing kind may not pose a danger to health for a decade or more, but the fast-growing kind can have serious implications for health within a few months. The slower prostate cancer is distributed rather evenly throughout the world, but the aggressive type of prostate cancer is more prevalent in developed countries where the consumption of animal products is much greater.[28]

A smaller-scale study has suggested another undesirable effect of eating meat. A study of 387 men born between 1949 and 1983 whose mothers consumed more than seven servings of beef per week found that 18 percent of them had less than 20 million sperm per liter, which is classified as subfertile by the World Health Organization. These men had difficulty conceiving a child with their partners as compared with those whose mothers ate less beef. Shanna Swan of the University of Rochester Medical Center in New York State has suggested that the hormones given to cattle to increase their growth, such as testosterone and progesterone, or the pesticides consumed by these animals might be to blame for the lower sperm counts.[29] For example, the hormones may interfere with fetal development of testes in the womb. These results have not yet been confirmed by a larger-scale study.

Harmful Organisms in Meat

There are additional risks associated with the consumption of meat because, being very rich in nutrients, meat provides a ready medium for the growth of pathogenic organisms. Some of these microorganisms may be picked up from the environment, while others may already be present in the carcasses of animals. Foodborne diseases cause 76 million illnesses, 325,000 hospitalizations, and 5,000 deaths in the United States each year. Although the pathogen that causes sickness cannot always be identified, 14 million illnesses, 60,000 hospitalizations, and 1,800 deaths have been attributed to the bacterial species that are found in the intestines of livestock, namely, *Campylobacter, Escherichia coli* O157:H7, *Listeria*, and *Salmonella*. It is likely that many other illnesses for which the source cannot be traced also have their origin in the livestock industry. *Salmonella* alone is responsible for 1.4 million cases of food poisoning a year and 400 associated deaths.[30] *Campylobacter* can lead to meningitis, arthritis, and neurological disorders; about 3.4 million people are sickened and 700 people die each year from this strain of bacteria. These pathogens infect the human population either by contaminating meat and dairy products or by transmission through the waste of farm animals. Paul S. Mead of the CDC has estimated the overall cost of these infections to the U.S. economy to be about $7 billion.[31]

The vulnerability of meat and other animal products to invasion by pathogens results in numerous recalls of these items from the market. While some products are recalled due to evidence of irregularities before the pathogens cause sicknesses in humans, many recalls occur as a result of disease. More than 50 products were taken off the market during both 2007 and 2008.[32] The most common cause for the recall of beef products was contamination with *E. coli* O157:H7, while contamination with the pathogens *Listeria* and *Salmonella* led to the recall of chicken products, sausages, and cheese. Despite monitoring by health authorities, the number of recalls and the amount of products that have been taken off the market have not shown any decline and may even be increasing. In 2007, there were 21 nationwide recalls of beef that removed 33.4 million pounds from the market because of possible *E. coli* O157:H7 contamination.[33]

A recall of unusually large magnitude occurred in February 2008, when 143 million pounds of beef that originated in a slaughterhouse in Southern California were taken off the market. The reason for the recall was that workers allowed a number of cows that could not stand—the so-called downers—to be slaughtered without contacting the veterinarian to investigate any evidence of mad cow disease. The enormous danger associated with this disease led to this massive recall. A few recalls each year occur when cow body parts not allowed in human food for fear of spreading mad cow disease, such as meat from the heads of cows, are mixed into supplies that are sent to market. Meats of foreign origin have also been subjected to a number of recalls. For example, the USDA Food Safety Inspection Service recalled beef from three different suppliers in Nicaragua in August 2008.[34]

Handling the meat of chickens at home before cooking poses a serious health problem due to the presence of the pathogens *Salmonella* and *Campylobacter* in this meat. A detailed study conducted by the periodical *Consumer Reports* in 2007 found that 83 percent of whole broilers bought in stores were infected with one of these two pathogens. *Campylobacter* was present in 81 percent of the broilers and *Salmonella* in 15 percent, and both organisms were present in 13 percent of the broilers. Thus, only 17 percent were free from both of these types of bacteria.[35] Chickens become infected with these organisms in the filthy housing that is used over and over again to raise up to 100,000 chickens at a time, and infection spreads rapidly through the flock in the closely packed environments. These organisms thrive in the birds' intestines and do not particularly harm the birds. When a bird is slaughtered, bacteria in its digestive tract wind up in the carcass. Both of these organisms can cause gastrointestinal illnesses in humans. Due to the overuse of antibiotics, many of the bacteria that contaminate chickens are resistant to drugs commonly used to cure bacterial infections in humans.

A VEGETARIAN DIET AND HEALTH

There is an important difference between a diet that has meat as a major component and one that consists entirely of plant products. Since meat provides an adequate supply of energy and nutrients in one package, a meat-based meal tends to have fewer components than a vegetarian meal. Foods derived from plants contain a large variety of antioxidants, dietary fibers, carbohydrates, and other phytochemicals (from the Greek *phyto*, meaning plants) that are beneficial for health. In addition to being good sources of many vitamins, minerals, and fibers, most plant products are low in fats and calories.

Items in a vegetarian diet may be divided into three categories: fruits and vegetables; grains, pulses, and legumes; and nuts and seeds. In addition, herbs and spices play a more important role in vegetarian diets than in meat-based meals. The number of phytochemicals is so large that it is difficult to assign an exact function to each one. Assigning a beneficial role to a component is further complicated by the fact, supported by several epidemiological studies, that many constituents of a vegetarian diet interact with each other in a synergistic manner to promote good health.

Numerous controlled studies have shown that a daily consumption of fruits, vegetables, grains, legumes, and nuts provides a level of protection against all chronic diseases. According to the American Institute for Cancer Research, the consumption of 400 grams or more per day of a variety of fruits and vegetables could, without any other lifestyle change, decrease overall cancer incidence by at least 20 percent. The evidence ranges from convincing to highly probable that diets high in vegetables and fruits protect against cancer of the mouth and pharynx, esophagus, lung, stomach, colon and rectum, larynx, pancreas, breast, and bladder.[36] A few studies have indicated that frequent consumption of these items also offers a level of protection from neurodegenerative diseases.[37]

Fruits and Vegetables: Antioxidants

The large number of phytochemicals in fruits and vegetables include dietary fiber, complex carbohydrates, beneficial oils, numerous substances that act as antioxidants, and many others that have not yet been characterized. Since plants are continuously exposed to the damaging effects of the sun's ultraviolet rays, they have developed numerous antioxidants to provide them with a certain level of protection. These substances are present in fruits, flowers, seeds, and in some cases, leaves as well. Of all the phytochemicals in vegetable products, the role of antioxidants is best understood. Antioxidants that are able to enter the bloodstream may neutralize the free radicals present in the blood, thus preventing the formation of plaques that may be formed by the oxidation of LDL cholesterol, and they also reduce incidents of oxidative damages to DNA, neural cells, and other components.

Although most vegetable products contain antioxidants, the amount and type of these chemicals varies. Two of the simplest antioxidants are vitamins C and E. While vitamin C is water soluble, vitamin E is a fat-soluble antioxidant. An important difference between water-soluble and fat-soluble compounds is that the fat-soluble compounds tend to stay in the body for longer periods and need not be consumed on a regular basis.

Another class of chemicals that humans obtain from fruits and vegetables in the diet are carotenoids, most of which are antioxidants. Various carotenoids impart a yellow, orange, red, or green color to food items and are found in abundance in many fruits and vegetables, such as mangoes, citrus fruits, peaches, pineapples, tomatoes, strawberries, apricots, guavas, watermelon, cantaloupe, carrots, pumpkins, and dark green leafy vegetables. In all cases, the amount of antioxidant pigment will be greater in items that have a deep and intense color. The most common carotenoids are alpha-carotene and beta-carotene, which can be converted to vitamin A in the body, and lycopene. Vitamin A is essential for normal growth and development and proper functioning of the immune system and helps to maintain vision. In many cases, chopping, pureeing, and cooking vegetables releases the carotenoids, which are then more easily absorbed by the body, particularly in the presence of a small amount of oil. The most common sources of lycopene are tomatoes and watermelon. Cooking tomatoes with small amounts of oil releases lycopene and increases its availability in the human body.

The most important phytochemcials that are powerful antioxidants belong to a class of compounds known as polyphenols; their molecules are made up of a number of small, ringlike structures known as phenols. The arrangement of phenolic rings and the number and location of side chains in each molecule distinguishes one type of polyphenol from another. Small structural differences in the molecules of polyphenols give rise to substantial differences in their properties, including variations in antioxidant activities. Due to subtle variations in structure, the number of polyphenols is very large; more than 4,000 such compounds have been identified in various fruits and vegetables. These chemicals provide a number of benefits to the plants—in addition to their antioxidant properties, the colors of polyphenols make them effective sunscreens, and their colors and aromas attract insects that help in pollinating and spreading the seeds. Polyphenols are divided into subclasses based on differences in their characteristics and origin. Flavonoids are found in most fruits and vegetables and also in tea and soybeans. Anthocyanins, a subclass of flavonoids, are powerful water-soluble antioxidants that impart a blue, purple, or blackish tint to berries, including strawberries, cherries, cranberries, raspberries, grapes, and black currants. In general, berries have the highest concentration of many types of powerful antioxidants.

To be effective, antioxidants in the diet have to survive the digestive process and enter the bloodstream, where they may neutralize the free radicals in the blood. The beneficial effects of some antioxidants have been established through large-scale epidemiological studies of human populations, which have shown that they provide significant protection against lung cancer[38] and CVD.[39] In a prospective study that followed the eating habits of more than 47,000 health professionals for eight years, it was found that those with the highest lycopene intake had a 21 percent lower risk of prostate cancer than those with the lowest intake.[40] Other studies have found a similar effect of lycopene, with somewhat different degrees of protection. A number of epidemiological studies have found that dietary intake of flavonoids reduces the incidence of certain types of cancers.[41]

While long-term epidemiological studies include some uncertainties due to variations in the daily routines of the participants, direct evidence of the effect of antioxidants can be obtained through studies in which cancer is induced in animals such as rats, mice, and hamsters by administering potent chemical carcinogens; afterwards, specific plant extracts are injected, and their effect on the development of tumors is monitored. The great variety of antioxidants in plant products means that studies of only a few flavonoids have been conducted, but these few studies have shown flavonoids to provide significant protection against cancers of the lung, oral cavity, stomach, esophageal system, stomach, colon, skin, prostate gland, and breast.[42] Cancers that were induced in animals by administering potent carcinogens either regressed or failed to develop when antioxidants were given to the animals.

Fruits and Vegetables: Other Benefits

In addition to antioxidants, fruits and vegetables contain a number of other components that have beneficial effects on human health. Epidemiological studies have shown that dietary fibers, trace minerals, antioxidants, and other chemicals in plant products interact with each other and the human body to provide some protection from degenerative diseases, including CVD[43] and cancer.[44]

Another molecule of interest is chlorophyll, the pigment that collects light from the sun and makes it available to plants for photosynthesis and is present in all types of vegetation. It has been suggested that the properties of this molecule to bind to some toxic chemicals may be helpful in promoting health. As mentioned earlier, cooking meat, particularly at high temperatures, produces known carcinogens such as heterocyclic amines and polycyclic hydrocarbons. Chlorophyll is known to bind strongly with these molecules to form a complex that is not broken down in the digestive system and hence passes through the body and is eliminated with the waste. A number of studies have looked at the protective effect of chlorophyll against aflatoxin, a powerful carcinogen that causes liver cancer and is produced when molds begin to develop on rice and some seeds. Aflatoxin is often used in studies to induce cancer in animals so that strategies for combating the effects of aflatoxin may be developed. Chlorophyll combines with aflatoxin to make a stable compound that is eliminated from the body, thereby sparing the body from the carcinogenic activities of this molecule.

A number of studies have shown that the benefits of phytochemicals are not limited to their antioxidant properties and that they help in reducing the incidence of chronic disease through other mechanisms. Two processes that provide protection against chronic diseases are the regulation of cells in the body and the prevention of inflammation. Cells continuously communicate with each other by excreting small molecules that act as messengers, or signals, for other cells. These signaling molecules help in the growth, development, and repair of cells and in the elimination of cells that have become defective because of mutations. There is some evidence that flavonoids in the bloodstream may regulate these processes to facilitate the development of healthy cells and silence genes that might otherwise foster disease. The concentration of flavonoids needed to control cell-signaling pathways is expected to be much lower than that needed to prevent damage done by free radicals.

Inflammation has been implicated in the development of many chronic diseases, including CVD, Alzheimer disease, and certain types of cancer. Inflammation is a precursor to cardiovascular disease because it provides a site for the buildup of arterial plaque and deposits of oxidized cholesterol. Inflammation damages the collagen in arteries so that, over time, tissues begin to scar and become less elastic. Blood vessels in these tissues become inflexible, and blood leaks into surrounding tissues. The vicious process continues, leading to atherosclerosis with deposits of oxidized cholesterol. The final step that leads to the formation of a clot is the aggregation of platelets circulating in the blood around the inflammation site.

A number of animal models and clinical studies strongly suggest that inflammation also plays an important role in chain reactions that ultimately lead to Alzheimer disease. Normal brain cells are disrupted as a result of inflammation that causes proteins in the brain to misfold.[45] Chronic inflammation has also been implicated in the growth and development of many types of cancers.[46] Flavonoids found in fruits and vegetables have been shown to have anti-inflammatory properties in laboratory studies; thus they may prevent or delay the onset of some chronic diseases. Juices of fruits and vegetables have been found to play an important role in delaying the onset of Alzheimer disease.[47]

Sugar molecules circulating in blood also act as oxidants. As a result of oxidation, collagen proteins become linked with sugars, resulting in damage to the microscopic blood vessels that cause most complications of diabetes. Damaged blood vessels allow large bloodborne molecules to migrate out of the bloodstream and accumulate between the surrounding tissues, causing chronic inflammation. Antioxidants found in berries may protect us from this type of cellular damage and thus reduce complications in people with diabetes. There is some evidence that, in addition to fighting inflammation, heart disease, and cancer, these chemicals can counter obesity and elevated blood sugar. Polyphenols in berries reduce blood glucose levels after the ingestion of starch-rich meals and may be helpful in regulating blood sugar in people with type 2 diabetes.[48]

Since there is usually a significant delay between harvesting the produce from plants and consuming them, it is important to know whether the antioxidants in fruits and vegetables survive the processes of storage, freezing, and cooking. Although it is impossible to collect such data for all chemicals in plant products, research has shown that many flavonoids are not degraded by freezing, processing, and cooking. In fact, in some cases, antioxidant activity is increased during processing due to the release of additional phenolic compounds.[49] However, it is not known whether all flavonoids are unaffected by processing. The more antioxidants that are absorbed by the body and become incorporated in the bloodstream, the greater the beneficial effects. How much is absorbed depends on the characteristics of the compound and the bacterial colony in the intestines. In general, isoflavones from soybeans and legumes are easily absorbed by the body. The absorption rate of anthocyanins in berries and grapes is also very high.[50]

Although many studies have been carried out to examine the effect of a diet rich in both fruits and vegetables, the mechanisms through which these two classes of agricultural products provide benefits may be significantly different. Their roles may not be interchangeable; instead, they may play complementary roles in providing protection. Professor Martha Morris of Rush University Medical Center studied the eating habits and cognitive decline of 3,718 older residents of Chicago and discovered that those who ate more than 2.8 servings of vegetables daily, particularly green leafy vegetables, had a 40 percent slower rate of cognitive decline with aging. Her conclusion was that vegetables are even more effective than fruits in slowing the rate of cognitive changes in adults.[51]

Cruciferous vegetables—broccoli, cauliflower, cabbage and brussels sprouts—have received particular attention because of their effects on the incidence of some types of cancers. These vegetables contain molecules known as sulforophanes. A number of studies have shown that regularly consuming more than three servings per week of these vegetables reduces the risk of developing bladder, prostate, and lung cancers.[52]

Grains and Legumes

The most common cereals used in foods are wheat, rice, corn, oats, and rye. Legumes that are normally consumed by humans include beans, peas, lentils, soybeans, and peanuts. Cereals and legumes are important sources of proteins, carbohydrates, dietary fiber, and many other useful chemicals. An intact grain has an outer layer called bran, a germ that contains the genetic information for the future seedling, and a comparatively large sac known as the endosperm. The outer bran layer contains fiber, B vitamins, 50 to 80 percent of the grain's minerals, and other helpful chemicals. The germ contains niacin, vitamin B6, vitamin E, and unsaturated fats. The endosperm, commonly associated with the grain, contains a large amount of simple and some complex carbohydrates and comparatively smaller amounts of protein and B vitamins. The bran and germ, although much smaller than the endosperm, contain most of the protein and other healthful chemicals in the grain.

In the past and in some cultures continuing today, grains were typically consumed either in their whole, intact form, or as coarse flours produced by stone grinding. Processing of grains with the modern high-power machines produces fine flours with very small particles. Milling also removes the outer bran layer and much of the germ. The resulting refined grain products contain more starch but lose a substantial amount of the dietary fiber, vitamins, minerals, essential fatty acids, and other beneficial components. Because of the loss of bran and the pulverization of the endosperm, refined grains are digested and absorbed more rapidly than whole-grain products and tend to cause more rapid and larger increases in the levels of blood glucose and insulin.[53]

Cereals such as wheat, corn, and rice have a protein content of between 7 percent and 14 percent, which is greater by a factor of 5 than that in potatoes and an order of magnitude greater than that in most vegetables. Legumes contain about 20 percent protein, with the exception of soybeans, which have a protein content of around 40 percent. All proteins are made of 20 building blocks known as amino acids, 8 of which are known as essential amino acids because they cannot be synthesized by the body and must be consumed on a regular basis. In addition, infants and growing children require four sulfur-containing amino acids in their diet for the healthy development of their bodies.

Unlike animal products, which have essential amino acids in the same proportion as that required by the body, both legumes and cereals are deficient in some essential amino acids. However, their compositions are complementary, so that combining them in a diet provides all the necessary amino acids. One problem often attributed to beans and legumes, especially legumes, is that they cause flatulence, because humans do not have the enzyme (alpha-galactosidase) that is required to rapidly digest them. Traditional cultures consumed cereals and legumes together in most meals to take advantage of the complementarities of their amino-acid composition. Some legumes, such as soybeans, chickpeas, and lentils, contain enzyme inhibitors that may interfere with the digestion and assimilation of proteins in the body. For this reason, they are known as antinutritive factors. Traditional cultures developed methods of prolonged soaking and extended cooking that substantially reduce the effect of these antinutritive factors.

Whole grains are concentrated sources of dietary fiber, vitamins, minerals, and complex carbohydrates known as oligosaccharides. Dietary fiber refers to plant cell wall components, which contain minerals, antioxidants, and other healthful chemicals. Fiber and carbohydrates that are not digested by enzymes in the small intestine are fermented in the large intestine (colon) by the resident anaerobic bacteria and other microorganisms. This process produces useful short-chain fatty acids that are absorbed by the body and provide building blocks for various cellular components. Soluble dietary fiber, found in legumes such as beans and lentils, vegetables, fruits, and oat bran, is easily degraded by bacteria in the colon. Insoluble fiber is found in whole grains, flaxseed, and some vegetables, such as celery and carrots. Insoluble fiber controls and balances the acidity (pH) of the intestine and reduces the accumulation of toxic waste in the colon. Most plant products contain both soluble and insoluble dietary fiber in different proportions. Oats, rye, and barley contain about one-third soluble fiber and two-thirds insoluble fiber. Wheat is lower in insoluble fiber content than most other grains, and rice contains virtually no insoluble fiber.

Among the legumes, soybean-based foods have received particular attention, mainly because of their very high protein content, almost double that of meats and other legumes. In addition to the useful chemicals that are found in all pulses, soybeans contain an estrogen-like compound known as isoflavone that mimics the properties of the hormone estrogen to some extent. The major tissues targeted by plant-based estrogens are the reproductive organs (uterus, breast, and prostate) and the cardiovascular tissues. A few studies indicate that a daily intake of isoflavones lowers the risk of heart disease, some types of cancer, and type 2 diabetes.[54]

Another healthful plant product is flaxseed, a small, greenish brown seed with a hard coating. While the human digestive system cannot break flaxseed down in its intact form to make use of its ingredients, flaxseed that is freshly ground is an excellent source of polyunsaturated omega-3 fatty acids, dietary fiber, and molecules known as lignans. These molecules are metabolized in the digestive tract to form phytoestrogens that have properties similar to the estrogenlike molecules in soybeans. However, flaxseed is a more potent source of phytoestrogens than soybean products, in addition to being an excellent source of dietary fiber. Flaxseed has received particular attention because it is an excellent source of omega-3 fatty acids and is also useful in preventing some types of cancers.

In some animal studies, cancer was induced in mice by using chemical carcinogens, and the influence of flaxseeds on the growth and development of their cancer was monitored with appropriate controls. It was found that including 10 percent flaxseed in the diet of mice reduced the growth of tumors, both in number and size, and prevented metastasis. Flaxseed was also found to prevent the metastasis of melanoma in mice that had been injected with melanoma cells.[55] Because of the advantageous effects of flaxseeds in animal studies, a pilot study was conducted of 25 men who were scheduled for surgery for prostate cancer. They were instructed to eat a low-fat diet supplemented with 30 grams (about 1 ounce) of ground flaxseeds daily. There was an improvement in all markers associated with prostate cancer, including PSA, total testosterone, and total serum cholesterol.[56]

A number of large-scale studies have shown that a regular intake of whole grains and legumes significantly reduces the risk of cardiovascular disease, colon cancer, and type 2 diabetes.[57] Other epidemiological studies have found that regular intake of whole grains alone is protective against some types of cancers, CVD, diabetes, and obesity. There may be many mechanisms that interact to provide this protection, since whole grains contain a variety of nutrients and other useful compounds. An improvement in the environment of the gut is beneficial for the immune system overall.

Table 7.3. Effect of food intake and the risk of cancer

Type of cancer	*Dietary factors that lower risk*	*Estimated range of lowering risk (min–max)*	*Factors that increase risk*
Lung	Fruits	20%–33%	Beta-carotene supplements
Stomach	Fruits and vegetables	66%–75%	Salty foods
Breast	Vegetables	33%–50%	Alcohol, obesity
Colon, rectum	Dietary fiber, garlic	66%–75%	Red or processed meat, alcohol
Mouth, pharynx	Nonstarchy vegetables, fruits	33%–50%	Alcohol, salted fish
Liver	—	—	Alcohol, contaminated food
Cervix	Fruits and vegetables	33%–50%	—
Esophagus	Nonstarchy vegetables, fruits	50%–75%	Alcohol, dietary deficiencies
Prostate	Foods containing lycopene or selenium	10%–20%	Dairy fats, meats

Source: American Institute for Cancer Research, www.dietandcancerreport.org/?p=er&J, reprinted with permission.

Whole grains are also rich in antioxidants, trace minerals, and phenolic compounds that have been linked to the prevention of chronic diseases. Additionally, unlike some other foods, whole grains and legumes do not lead to a rapid increase in the levels of sugar and insulin in the blood, and so they do not contribute to diabetes and obesity in the ways that some other foods do.

In contrast to the complex carbohydrates found in whole grains, the simple carbohydrates in refined flour increase the risks of cancers of the colon and breast. Although most nutritional studies are conducted with isolated nutrients to determine their specific role in the prevention of disease, a few studies have followed the effect of consuming whole grains on overall health. It has been found that regularly consuming whole grains provides the benefits associated with antioxidants and leads to an improvement in metabolic syndromes.[58] Eating foods with low glycemic indexes also delays the return of hunger by increasing the sensation of fullness and hence may be used as part of a weight-reducing diet.

A number of factors may contribute to the beneficial effects of eating whole grains and legumes. Unlike meat and dairy products, which do not contain any fiber at all, whole grains and dry beans are rich in soluble fiber, complex carbohydrates, folate, magnesium, and other minerals. Potassium in fiber may decrease the incidence of cardiovascular diseases by lowering blood pressure. Soluble fiber establishes strong bonds with dietary cholesterol, thus preventing its absorption by the body. It also moderates the level of glucose in the blood. The benefits of insoluble fiber include a better intestinal environment that reduces the risk and occurrence of colorectal cancer, hemorrhoids, and constipation. Although wheat fiber has not been found to lower serum cholesterol levels, numerous clinical studies have demonstrated that increasing the intake of oat fiber results in modest reductions in the levels of total and LDL-cholesterol. While meat and dairy products have a large proportion of saturated fats, the small quantities of fats in legumes are mostly unsaturated.

Nuts and Seeds

Nuts include almonds, cashews, pecans, pistachios, walnuts, hazelnuts, macadamia nuts, pine nuts, and Brazil nuts (although many of these are not true nuts in the botanical sense). Although peanuts are a legume, they are nutritionally similar to nuts and are usually included in this list. A few years ago, nuts were considered unhealthy because of their high fat content, typically 150 to 200 kcal per ounce. However, the fats in nuts are of the healthier monounsaturated and polyunsaturated kind, and consuming nuts in limited amount is now considered to be an essential part of a healthy diet. Macadamia nuts have only monounsaturated fatty acids (and therefore are not as healthy as other nuts), whereas walnuts, cashews, peanuts, pecans, and pistachios contain significant proportions of polyunsaturated and monounsaturated fatty acids. While animal cells necessarily contain cholesterol to maintain the structural integrity of membranes, the membranes of the cells of plants contain compounds known as phytosterols that are structurally somewhat similar to cholesterol. Nuts, in general, are rich in phytosterols. When phytosterols are ingested, they interfere with the synthesis of cholesterol; studies have shown that including phytosterols in the diet lowers serum cholesterol. Although the mechanism by which nuts provide health benefits may be complicated and may involve an interaction between various components, their beneficial effects have been established by a number of epidemiological studies.

The effect of regular consumption of nuts on cardiovascular health has been studied by a number of groups and has consistently been shown to significantly reduce the risk of coronary heart disease. Four large prospective studies found that regularly eating nuts has a beneficial effect on serum lipid and cholesterol, important markers of cardiovascular health.[59] A 14-year study of more than 86,000 women participants in the Nurses' Health Study found that those who consumed more than five ounces of nuts at least twice weekly had a risk of coronary heart disease that was 35 percent lower than those who ate less than one ounce of nuts each month.[60] Another study, which followed 21,000 male health professionals over a period of 17 years, discovered that those who consumed nuts at least twice a week had a 53 percent lower risk of sudden cardiac death than those who never or rarely consumed nuts.[61] A few studies have found that in addition to providing some protection against cardiovascular diseases, a regular intake of nuts or peanut butter provides some protection against diabetes.[62] The available cumulative data do not indicate that people who eat nuts on a regular basis have a higher body mass index or a tendency to gain weight.

Herbs and Spices

Spices are dried aromatic parts of plants, generally consisting of seeds, berries, roots, pods, and leaves. When added to foods, they can have a number of complementary and overlapping effects. In addition to being strong antioxidants, many spices activate detoxifying enzymes, reduce inflammation, and have antibacterial and antiviral effects. The antioxidant activities of cloves, cinnamon, pepper, ginger, and garlic remain intact even after boiling for 30 minutes, suggesting that antioxidants in these spices are very stable. A preliminary in vitro (in a petri dish) screening of 35 different spices and herbs indicated that cloves, cinnamon, chili, horseradish, cumin, tamarind, black cumin, pomegranate seeds, nutmeg, garlic, onions, and bay leaf help to kill microbes such as *Bacillus subtilis, E. coli,* and *Saccharomyces cerevisiae.*[63] Extracts of several commonly used Indian spices (see below) have been shown to inhibit oxidation of fats and cholesterol. Combinations of spices have a synergistic effect that is exploited in Indian cooking by using them together.

Timothy Bates of the University of Nottingham applied capsicum, the compound that makes chili hot, to cancerous human lung and pancreatic cells. He found that capsicum is effective in killing cancerous cells while leaving the noncancerous cells intact. Vanilloids, the family of molecules to which capsicum belongs, bind to proteins in the mitochondria of cancerous cells and trigger apoptosis (cell death).[64] The antioxidants in capsicum quench free radicals, and vanilloids prevent the promotion of tumors by arresting the cycle that leads to the proliferation of cancerous cells.

That capsicum and other spices have been used in food preparation for thousands of years shows that they are safe to eat and have minimal side effects, and Bates suggests that cancer patients or those at risk of developing cancer could be advised to eat a diet that is richer in spicy foods. Cancer Research UK maintains that eating vast quantities of chili pepper is not advisable but that the risk of cancer can be reduced by eating a healthy balanced diet with plenty of fruits and vegetables. Several other spices and their constituents have also been found to prevent the growth of cancerous cells. Extracts of saffron have been shown to inhibit and retard the formation of skin, colon, and soft tissue tumors, and ginger has been found to retard the growth of some types of tumors.

A polyphenol named curcumin is responsible for the yellow color of the spice turmeric, an essential ingredient of Indian curries that is also used as a food preservative and a traditional herb in India. The well-established anti-inflammatory properties of turmeric make it a useful compound to help prevent chronic diseases; it has also been shown to help prevent cancers of the skin, colon, oral cavity, and liver in mice. In another study, Sharon McKenna and her team at the Cork Cancer Research Center in the United Kingdom found that curcumin destroys cancerous cells in the throat within a very short time. The cells began to digest themselves after curcumin triggered lethal cell death signals. This opens up the possibility of treatment of esophageal cancer with curcumin.[65]

Through an epidemiological study, Tze-Pin Ng and colleagues at the National University of Singapore found that curcumin inhibits the buildup of amyloid plaques in people with Alzheimer disease, improving their cognitive abilities.[66] Studies with mice have also shown that curcumin reduces cognitive decay and delays the onset of Alzheimer disease.[67] The common and frequent consumption of turmeric in India may contribute to the low incidence of Alzheimer disease in that country. The brain is highly susceptible to oxidative damage because of its high metabolic load and an abundance of oxidizable material such as polyunsaturated fatty acids that form the plasma membranes of neural cells. Curcumin is a strong antioxidant that seems to protect the brain from oxidation of fats and neutralize some free radicals.[68]

Cuisines that traditionally do not include much meat use a wide variety of spices for seasoning. Since spices are often used in greater numbers and amounts in vegetarian cooking, they act in conjunction with the phytochemicals in vegetable products to provide numerous health benefits, including modulation of hormonal activity and metabolism and inhibition of inflammation. A number of studies have indicated the benefits of spices for ailments other than chronic diseases. Some spices have been found to relieve and prevent cluster headaches, sinusitis, muscle pain, psoriasis, and ulcers.[69]

FATS AND OILS

There are many types of fats and oils of plant and animal origin. We often refer to fats and oils simply as "fats." Fats, or, more accurately, fatty acids, are essential nutrients for the human body. Consumed in moderation, fats help to keep the body in good condition.

Fats and oils, with common sources.

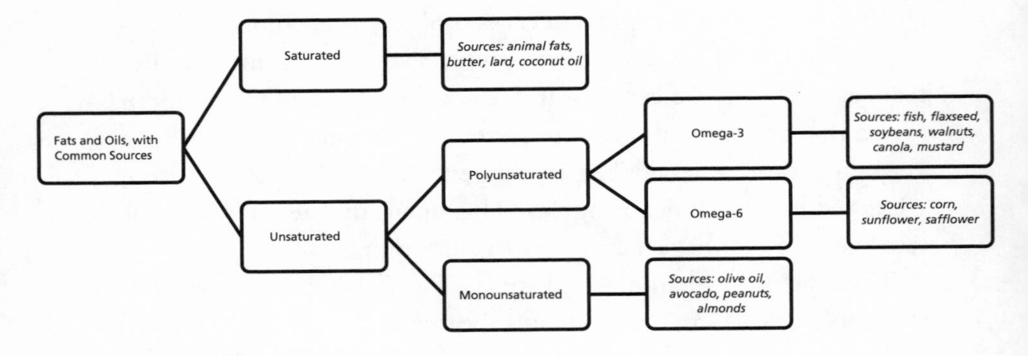

They are the building blocks of cell membranes and also provide energy to various organs of the body. Consuming too much fat or the wrong kinds of fat, however, can be a major factor contributing to chronic diseases.

A typical molecule of fat consists primarily of a long chain of carbon atoms with atoms of hydrogen attached to each carbon in the chain. The length of the chain and the arrangement of attached hydrogen atoms impart a distinctive characteristic to each fat. Saturated fats are solid at room temperature and have long shelf lives. The term "saturation" means that the substance cannot absorb any more hydrogen; carbon atoms in each molecule have hydrogen atoms attached to the maximum capacity. Unsaturated fats or oils are liquid at room temperature and may be spoiled by exposure to heat, light, and air. Unsaturated fatty acids may be divided into two groups—monounsaturated (MUFAs) and polyunsaturated (PUFAs), depending on whether there is only one site in the molecule without a hydrogen atom or many such sites. Unsaturated oils can be made partially saturated by bubbling hydrogen through them at high temperatures under suitable conditions, as is commercially done to increase the shelf life of fats in products such as margarine and some cookies. However, some of the synthetically added hydrogen atoms form a slightly different configuration than in natural fats and form the so-called trans fats. Numerous studies have shown that trans fatty acids tend to increase the level of LDL cholesterol and lower the level of HDL cholesterol, and in this way they increase the risk of CVD.

Polyunsaturated fats and oils are known as essential fatty acids because they are not synthesized within the body and must be obtained by consuming fats of the omega-3 and omega-6 variety (also known as n-3 and n-6, respectively, based on some structural differences in their molecules). To perform a number of metabolic processes and function properly, the body requires both kinds of PUFAs; polyunsaturated fats are involved in processes that support the immune and nervous systems and regulate heart beat and blood pressure. Omega-3 oils have received particular attention because of their health benefits.

The unsaturated vegetable oil found in most seeds is of the omega-6 variety; it includes 18 carbon atoms and is known as linoleic acid. A few nuts and vegetable products contain the alpha linolenic acid (ALA) that also has 18 carbon atoms but has the omega-3 configuration. While the majority of oils in seeds and vegetables are of the omega-6 kind, oils extracted from flaxseed, canola, soybean, and mustard contain significant amounts of omega-3 oils. Other vegetable sources of omega-3 oil include walnuts, beans, winter squash, kiwi fruit, avocado, and olive oil. Humans and other animals do not have the physiological mechanism to convert omega-6 oils to omega-3, or vice versa, after ingestion. Hence the amount of omega-3 oils in an animal product depends, to a large extent, on the feed given to it. This is true for meats from land and sea animals and also for eggs and milk. Eggs that are rich in omega-3 oils have been produced by feeding appropriate seeds to the laying hens.

The omega-3 oils in fish are different from those found in vegetable sources, because they have somewhat longer chains of carbon atoms—EPA (eicosapentaenoic acid) has 20 carbon atoms, and DHA (docosahexaenoic acid) has 22 carbon atoms. A number of epidemiological and clinical studies have established the benefits of these two oils in reducing the incidence of CVD,[70] bone loss, certain types of cancers,[71] and neurological disorders such as Alzheimer disease.[72] A large proportion of the DHA and EPA in the body is present in the brain, eyes, and heart tissues and therefore is considered to be necessary for the efficient functioning of these parts of the body on a cellular level. The cardiovascular benefits from long-chain omega-3 fatty acids, as documented in several prospective studies and randomized clinical trials, have mainly been attributed to their ability to prevent inflammation, atherosclerosis, and clotting of the blood. EPA is thought to reduce inflammation by modifying the immune response.

There are enzymes in the body that can convert the 18-carbon omega-3 ALA found in vegetable products to the longer EPA or DHA found in fish oil, but the efficiencies of these processes are very low. On the other hand, the body requires only small amounts of long-chain fatty acids, and the natural elongation process may be able to provide the required amount of these oils if sufficient amounts of plant-based omega-3 oil are included in the diet. Medical science has not established the daily requirement of EPA and DHA, but values on the order of one gram per day are generally considered to be sufficient.[73] A number of population studies have shown that plant-sourced ALA in itself has a protective effect against coronary disease, although it is likely that some of the ingested ALA is converted to DHA and EPA within the body. Several chronic diseases have been associated with diets deficient in omega-3 fatty acids.

A diet rich in omega-3 fatty acids is garnering appreciation for supporting cognitive processes and reducing the incidence of chronic diseases in humans. In contrast, epidemiological studies indicate that diets containing large amounts of trans and saturated fats adversely affect cognition. It seems likely that the very low ratio of omega-3 to omega-6 oils in the typical Western diet is partly responsible for the prevalence of chronic diseases. For this reason, medical authorities often recommend consuming oily fish or fish oil for maintaining good health. Although such suggestions have led to an increase in the consumption of fish, this trend is bound to face serious obstacles. As noted in earlier chapters of this book, the stock of wild fish is declining at a rapid rate, many of the edible species are in danger of being wiped out, and the populations of all species of fish are projected to fall precipitously within the next few decades. The production of fish in aquaculture facilities does not provide a large store of omega-3 oils, because farmed fish generally contain a smaller proportion of the desired kind of oil than wild fish do.[74]

A diet containing oily fish is often recommended by health practitioners for its health benefits, but fish are consumers and not producers of the beneficial oils since they, like humans and animals, are not able to synthesize oils within their body; they can only consume them for energetic needs and modify them to a limited extent. The original sources of long-chain omega-3 fatty acids are the marine algae and phytoplankton at the bottom of the food chain. These organisms have developed mechanisms to synthesize long-chain unsaturated fatty acids because these acids provide them with the flexibility and fluidity necessary to survive in cold water.

Recently, it was discovered that marine algae can be cultured industrially to provide "fish" oil, thus providing oil without sacrificing fish.[75] The production of DHA by microalgae has now been commercialized. Indoor facilities that grow suitable strains of microalgae can provide omega-3 fatty acids at a much greater rate than outdoor systems and are more reliable sources of long-chain omega-3 oils of consistent quality. Another approach that has shown sufficient promise is using molecular biology to engineer soybeans or other plants so they will contain EPA and DHA in their seeds. In the future, both of these methods may be used to produce desirable oils in sufficient quantities in an inexpensive and reproducible manner.

Taking supplements of synthetically produced fish oil has a number of advantages over eating fish on a regular basis. For one thing, the concentration of the desired type of oil in any stock of fish shows significant variation, which means that you don't know how much omega-3 oil you are obtaining when you eat fish, as you do when you take supplements. Since the beneficial oils move up the food chain in a somewhat selective manner from their original source—the algae and phytoplankton—the amount present in fish raised in aquafarms depends on the feed given to them. Wild fish are also exposed to environmental contaminants such as methylmercury, PCBs, and dioxins. PCBs and methylmercury have long half-lives in the body and can accumulate in people who frequently eat contaminated fish, with serious health consequences. In contrast, oils derived from microalgae or soybean plants are not contaminated with these chemicals, and, being plant products, do not have any cholesterol. A commercial operation of this kind that produces omega-3 oils may be very efficient because it may produce about a hundred times more long-chain oil in a given area than aquaculture. Finally, the dwindling stock of wild fish will continue to increase the cost of obtaining oils from them, making it possible for only a small portion of the population to obtain fish oil from wild fish.

During the last few decades several epidemiological and interventional studies have been conducted to see how consuming various types of fats affects the incidence of chronic diseases. It is widely recognized that mortality and morbidity due to chronic diseases are lower in vegetarians than in omnivores.[76] There is a considerable difference between the type and amount of fat in these two types of diets. Vegetarian diets are slightly lower in total fat than omnivorous diets. In addition, vegetarians eat about one-third less saturated fat and about one-half as much cholesterol as omnivores.[77] Studies have also established that eating the right kind of fats in limited proportion is more beneficial than avoiding them altogether.[78] There appears to be no benefit in terms of cancer prevention from reducing fat intake from vegetable sources, and in the case of breast cancer, there is some suggestive but preliminary evidence that olive oil and other sources of monounsaturated fatty acids may modestly decrease the risk.[79]

Walter Willett of the Harvard School of Public Health examined the eating habits and incidence of chronic disease in 250,000 men and women for 30 years and concluded that the percentage of calories from fat or carbohydrates does not affect the occurrence of heart attacks, various cancers, and stroke, but that the type of fat or carbohydrate is important, with trans fats being particularly harmful and unsaturated fats having a beneficial effect. His research also showed that the intake of refined starch and sugar is related to a greater risk of heart disease and diabetes.[80]

GENERAL CONSIDERATIONS FOR A VEGETARIAN DIET

Epidemiological and clinical research during the last few years have provided us with a better understanding of the role that various food items play in facilitating or preventing the onset of chronic diseases. Such work has clearly established the benefits of a variety of plant products in maintaining good health and in eliminating or delaying the incidence of chronic disease. Antioxidants, complex carbohydrates, dietary fiber, trace minerals, and numerous other plant products help in reducing the incidence of many types of cancers, cardiovascular diseases, neurological disorders, and diabetes. In one study, the frequency of heart attacks and strokes was found to be 24 percent lower in vegetarians than in nonvegetarians.[81] Mortality from coronary heart disease is lower in vegetarians than in nonvegetarians, and vegetarian diets have also been successful in slowing the development of ailments related to heart disease and improving people's quality of life.[82] A joint report by the World Cancer Research Fund and the American Institute for Cancer Research displayed convincing evidence that a diet high in fruits and vegetables reduces the incidence of various types of cancers in the population.[83]

The human gastrointestinal system is colonized by a host of different types of bacteria. These beneficial bacteria perform a number of functions that keep the body operating properly. They provide enzymes in the small intestine that help in the digestion of food, produce some B vitamins, and help strengthen the immune system in the gut, where components of the digested food are absorbed by the body. Some pathogenic bacteria may also inhabit the intestinal system and secrete toxins and harmful chemicals that can lead to various diseases. The balance of beneficial and potentially pathogenic bacteria in the intestines is dependent on diet.

A vegetarian diet changes the environment of the human intestinal tract to promote the growth of beneficial bacteria at the expense of harmful pathogens that may secrete toxins. Products in a vegetarian diet such as fiber and roughage promote the growth of beneficial bacteria, and a large population of beneficial bacteria decreases the growth of pathogenic organisms. The environment of the colon of vegetarians is significantly different from that of meat eaters in a way that may provide a number of benefits, including a reduction in the incidence of colorectal cancer.

Vitamin B12, considered to be the Achilles' heel of the vegetarian diet, is produced by bacteria that reside in the forestomach of ruminants from food rich in fiber, usually hay and forage. There is some indication that this synthesis can also take place in the colon and small intestine of humans if they consume sufficient amounts of fiber and if the bacterial colony in their gut has not been destroyed by the use of antibiotics.

It appears that the entire world is undergoing a nutritional transition toward a more energy-rich diet that includes substantial amounts of animal products. The consumption of animal products has been high in the developed world for at least a few decades and is still increasing at a slow rate. People living in developing countries who ate whole grains, beans, fruits, and vegetables as the major components of their diets are rapidly changing to follow the Western model. This trend is making a significant contribution to the widespread rise in obesity and chronic diseases. Many developing countries now face the double burden of undernutrition in the part of the population that does not have enough food and, in the affluent section of society, an increase in obesity-related chronic diseases due to energy-dense products containing large amounts of proteins, fats, and sugar. In countries like India, Mexico, Brazil, and Chile, the rates of diabetes, hypertension, and coronary heart disease were found to be consistently lower in rural areas where the traditional diet consists largely of vegetable products with small quantities of foods of animal origin and differs from the higher-income urban diet in the type and amount of fats, proteins, and carbohydrates.[84]

Recent studies have shown that an excessive intake of calories, in addition to causing obesity and contributing to metabolic syndromes such as high blood pressure, high cholesterol, and high blood glucose, negates the positive effects of certain diets. An abundance of food, particularly with a high-energy content, has deleterious effects on the brain that may lead to neurodegenerative diseases.[85] Judging by the increasing rates of obesity and Alzheimer disease in Western countries, excessive food intake in wealthy nations may be almost as harmful as food shortage in poor countries. In this context, it may be useful to note that several countries with limited resources have a reduced incidence of neurological disorders such as Alzheimer disease, indicating the association between these disorders and rich diets.

Foods of animal origin contain trace minerals such as iron, zinc, and phosphorus in a form that can be easily absorbed by the body. It is sometimes said that a vegetarian diet, even though it may contain these elements, will lead to their deficiency because they are not consumed in a bioavailable form; however, epidemiological studies in developed countries have not demonstrated any adverse health effect of the lack of these minerals in vegetarians,[86] and a moderately lower intake of iron has been hypothesized to reduce the risk of some chronic diseases. An excess of iron is known to create multiple problems, partly because the human body cannot easily remove it from the bloodstream. The relatively lower absorption of iron from vegetable products may be advantageous because the rate at which it is incorporated may be regulated according to bodily need; the body may absorb less when there is a large store of iron in the blood and absorb nearly as much iron as is absorbed by meat eaters when there is a deficiency of iron in the blood. While absorption of an adequate amount of iron is needed for proper body function, an excess of iron can have pathological effects. Data from animal studies indicate that, once in the bloodstream, ionic, or free, iron can promote the production of free radicals, contribute to atherosclerosis by oxidizing LDLs, and directly contribute to ischemic myocardial damage. Although vegetarian sources of zinc, such as legumes, whole grains, nuts, and seeds, do not produce zinc in a form that can be easily assimilated by the body, deficiencies can be avoided by consuming whole grains, because the fraction of zinc absorbed from unrefined foods is greater than that available from highly refined foods.

Whole Foods versus Supplements

To get the health benefits of fruits, vegetables, grains, nuts, and seeds, a vegetarian diet has to include a variety of these types of foods. The sheer number of phytochemicals present in any single item is very large, and the exact role of many of them is not properly understood and may never be established. For example, the antioxidant activity of fruits comes from a whole range of chemicals and not from a single one. Vitamin C, a known and potent antioxidant, can only account for less than 0.4 percent of the antioxidant activity of whole apples. Another important feature of a varied diet is that its constituents interact to produce a beneficial effect that may be greater than that from individual food items. This synergistic effect works not only within various components of a single fruit or vegetable but also between various produce items when they are combined in a meal.

It is common to extract the active ingredients from fruits and vegetables and put them in the form of a pill or capsule. While this approach looks attractive, it has many flaws. Scientific studies done on the benefits of taking the active ingredient in a concentrated form have shown limited advantages; in some cases, concentrated extracts have even been found to have a harmful effect. The first reason why pills may fail to have the same advantage as the food item from which they are extracted is that a number of phytochemicals act in concert to provide the beneficial effect, and the identified component, like the vitamin C in apples, noted above, may simply be one of the many beneficial components in the natural product. In addition, the beneficial components of foods are released slowly during the digestive process, giving the body a chance to absorb them at a rate that is in tune with human metabolism. The active ingredient in a concentrated form may provide an overload that is too much for the body to handle and, if incorporated, may even have undesirable effects.

Another advantage of consuming antioxidants and other active ingredients in food rather than as supplements is that the chemistry of these agents can be influenced by what is eaten along with them. For instance, the carotenoids tinting most vegetables preferentially dissolve in fats and oils. Unless these pigments are eaten along with fat, the amount of carotenoid absorbed by the body is rather small. Steven J. Schwartz of the Ohio State University showed that avocado, a high-fat fruit, facilitates uptake by the body of carotenoid in tomatoes.[87] Isolating the identified active ingredient and making it available in a bottle has the greatest benefits for the manufacturing companies, because pills have longer shelf lives than fruits and vegetables, and a single establishment can provide it to large parts of the country. In contrast, marketing of agricultural products is tedious and highly susceptible to local conditions and has a low profit margin. Taking a pill with the desirable ingredient is psychologically satisfying to consumers because it saves them the hassle of buying and ingesting a number of food items. But there are several compelling reasons why it is better to obtain antioxidants and other active ingredients from foods rather than from supplements. (Fish oil is an exception, particularly if the oil has been obtained from fish in pristine waters.)

A number of clinical trials have shown that isolated chemicals and dietary supplements do not have the same beneficial effects as whole grains, fruits, vegetables, and nuts. Most flavonoid researchers agree that people should get healthy doses of these chemicals from colorful foods, not dietary supplements.

The search for a nutritional pill—a magic bullet—may not only be futile but even injurious to health. Indeed, several studies have shown that a megadose of a flavonoid can trigger harmful effects, either because a very large amount of flavonoid in concentrated form may be toxic for the body, or because a compound that acts as an antioxidant in small amounts may even facilitate oxidation when ingested in a concentrated form. In early case-controlled studies it appeared that beta-carotene was a cancer-protective agent; however, randomized controlled trials of beta-carotene found that the nutrient, when isolated, was either neutral or increased the risk of lung cancer in smokers.[88] Homer Black of Naylor College of Medicine in Houston showed that a diet overly enriched with beta-carotene promoted skin cancer; the increased beta-carotene increased the oxidative damage from ultraviolet light rather than protecting against it. While epidemiological studies have shown an inverse relationship between *dietary* carotenoids and various types of cancers, three of four interventional trials using high-dose beta-carotene supplements did not show protective effects against cancer or CVD. Rather, the high-risk population of smokers and asbestos workers who took high-dose beta-carotene supplements had increased rates of cancer and angina. In another study, Edgar Miller of Johns Hopkins University analyzed the results of 19 clinical trials involving 135,967 patients and found that there were 39 additional deaths per 10,000 people who were taking vitamin E doses exceeding 400 international units per day.[89]

Vegetarianism and Weight Control

Long-term adherence to a vegetarian diet is likely to reduce the incidence of obesity. Animal products—meat, eggs, and dairy products—have an energy content that is greater than that of a typical vegetarian diet. Studies that establish the benefit of vegetarian diets require four to six servings each of grains, fruits, and vegetables daily, leaving little room for high-energy foods.

A survey conducted in the United States of 13,000 vegetarians over the age of 18 years and an equal number of people who ate meat found that vegetarians had a mean BMI of 23, significantly lower than the BMI of 26 for those who ate meat. The nutrition contents of the diets of the two groups were also significantly different. Vegetarians had a smaller but adequate amount of protein, more complex carbohydrates, and smaller amounts of fats in their diets. Compared with people who ate meat, vegetarians had a greater proportion of grains, legumes, nuts, and seeds. On the whole, the dietary pattern of people who practiced vegetarianism for long periods was healthier than the pattern for those who ate meat.[90]

A diet that is rich in whole grains and legumes is useful for controlling weight because it produces a feeling of satiety without overloading the body with a large amount of energy. Whether a person loses weight or not after switching to a vegetarian diet depends on which food items and how much of them are consumed. Even though following a healthy vegetarian diet for a long time is likely to keep a person's weight in control, vegetarianism should be considered to be part of a healthy lifestyle rather than a weight-loss program.

Importance of Variety in Vegetarianism

The healthy effects of a vegetarian diet depend on eating a variety of fruits and vegetables so that dietary fiber, flavonoids, antioxidants, and carotenoids—as well as chemicals whose functional role has not yet been established—can act in concert to provide health benefits. When people consume foods that have not been highly processed, and hence have intact nutritional components, it may not be possible to assign an exact role to each component, but such an approach ensures the built-in redundancy of multiple agents with independent, overlapping, and perhaps correlated mechanisms.[91] A growing body of scientific evidence indicates that wholesome vegetarian diets offer distinct advantages compared to diets containing meat and foods of animal origin. The large number of distinct phytochemicals in these items makes variety an essential requirement of a vegetarian lifestyle.

The American Dietetic Association states that well-balanced vegetarian diets are suitable for people in all stages of the life cycle, including children, adolescents, pregnant and lactating women, the elderly, and competitive athletes. Following is the association's 2009 statement:

> It is the position of the American Dietetic Association that appropriately planned vegetarian diets, including total vegetarian or vegan diets, are healthful, nutritionally adequate, and may provide health benefits in the prevention and treatment of certain diseases. Well-planned vegetarian diets are appropriate for individuals during all stages of the life cycle, including pregnancy, lactation, infancy, childhood, and adolescence, and for athletes. A vegetarian diet is defined as one that does not include meat (including fowl) or seafood, or products containing those foods . . . An evidence-based review showed that vegetarian diets can be nutritionally adequate in pregnancy and result in positive maternal and infant health outcomes. The results of an evidence-based review showed that a vegetarian diet is associated with a lower risk of death from ischemic heart disease. Vegetarians also appear to have lower low-density lipoprotein cholesterol levels, lower blood pressure, and lower rates of hypertension and type 2 diabetes than nonvegetarians. Furthermore, vegetarians tend to have a lower body mass index and lower overall cancer rates. Features of a vegetarian diet that may reduce risk of chronic disease include lower intakes of saturated fat and

cholesterol and higher intakes of fruits, vegetables, whole grains, nuts, soy products, fiber, and phytochemicals.[92]

For most people, vegetarian diets can help prevent or delay the onset of diseases such as CVD, hypertension, diabetes, cancer, osteoporosis, renal disease, and dementia, as well as diverticular disease, gallstone disease, and rheumatoid arthritis.[93] A wholesome vegetarian diet including a variety of foods should be adopted for its numerous health benefits alone—even without taking into consideration the known adverse effects of a meat-based diet.

8 * DAIRY

Milk is different from other animal products for several reasons. A cow can provide us with fresh milk on a daily basis for a number of years by eating grass and hay, as well as crop residues, household waste, and seeds whose oil has been extracted, none of which are of any direct use to humans. The indigestible fiber consisting of cellulose and hemicellulose in these feed items is converted into useful substances by the multitude of microorganisms living in the cow's rumen so that the ultimate product—milk—is very nutritious. In addition, unlike cattle raised for meat, a dairy cow does not have to be sacrificed to get its milk and remains productive for a few gestation cycles. How much milk a cow produces and how long a cow produces sufficient quantities of milk depends somewhat on its feed, its environment, and whether synthetic chemicals are administered.

There are about 240 million dairy cows in the world today. The cows are more or less evenly distributed between developed countries and underdeveloped and emerging economies, but superior genetic strains, optimized feed, and control of living conditions allows the dairy farms in the developed world to produce more than four times the milk produced by cows in developing countries. The trend toward the consumption of milk and dairy products is similar to that for meat—it is increasing rapidly in developing countries from very low values but will remain substantially lower in developing countries than in the developed world for the foreseeable future. The average per capita consumption of milk in developing countries is projected to increase from the present value of about 45 kilograms (11.9 gal) per year to 62 kilograms (16.4 gal) per year in 2020, still less than a third of the annual consumption of more than 200 kilograms (52.9 gal) in the developed world. Similar to the situation with meat, it is the emerging middle class in developing economies that will be the driving force behind much of this increase.

The total worldwide consumption of milk is expected to increase from 450 million tons to 654 million tons in 2020,[1] with a similar growth pattern in the years thereafter. Milk produced by the nine million cows in the United States accounted for about $27 billion of cash receipts for producers in 2005, making it the second largest agricultural commodity in the country.

MODERN DAIRY OPERATIONS

Modern dairy operations place their emphasis on getting the maximum amount of milk in the most economical way—and this focus has changed the industry significantly. Large dairy farms maintain a herd of a few thousand milking cows. Genetic selection and breeding techniques have been used to develop breeds of cows that mature early and produce large quantities of milk; they are fed high-energy feed consisting of large amounts of grains, which is designed to maximize the output of milk. (The total grain used in the feed of dairy cows worldwide is expected to increase from 650 million tons in 2005 to 928 million tons in 2020.) Cows on modern dairy farms produce an enormous amount of milk—as much as 10 gallons per day, at such a rapid rate that they may have to be milked up to seven times a day. This method of dairy farming is spreading to all parts of the world to fulfill the increasing demand for milk and dairy products.

The dairy farm industry in the United States produces 82.5 million tons of milk per year, making it the largest producer of milk in the world. While the national dairy herd has now shrunk to al-

Table 8.1. Milk production in selected countries, in millions of tons

Country	*Milk production*
USA	82.5
India	39.8
China	31.9
Russia	31.3
Germany	28.0
Brazil	25.7
France	24.2
New Zealand	15.0
UK	14.4
Ukraine	13.3
Poland	12.0
Italy	11.2
Netherlands	11.0
Mexico	10.4
Argentina	10.2
Turkey	10.0
Australia	9.5
Canada	7.9

Source: International Dairy Federation, www.fil-idf.org, with permission.

most half its size of 18 million in 1960, the total production of milk has greatly increased; per capita consumption has only slightly increased, while the population of the country has increased from 180 million to more than 300 million during this period. Selective breeding, improved technology, and the administration of growth hormones to cows has increased milk production per cow to three times the level of their forebears in the 1950s, and six times that of cows in 1900. As an extreme example of the lack of genetic diversity, most of the nine million dairy cows in the United States were sired by just a handful of bulls.[2] Americans drink more than 5 billion gallons of milk per year, and another 10 million gallons are used to make cheese. An average American drinks 22 gallons of milk and consumes 35 pounds of cheese per year. It takes roughly 10 pounds of milk (5 quarts) to make a pound of cheese; the exact amount depends on the type of cheese. The United States produces

Table 8.2. Per capita consumption of milk and cheese in selected countries

Country	*Liquid milk (liters)*	*Cheese (kg)*	*Total (milk equivalent, liters)*
Finland	183.9	19.1	375
Greece	69.2	28.9	358
Switzerland	112.5	22.2	335
France	92.2	23.9	331
Sweden	145.5	18.5	330
Netherlands	122.9	20.4	327
Germany	92.3	22.4	316
Italy	57.3	23.7	294
Norway	116.7	16.0	277
Austria	80.2	18.8	268
USA	83.9	16.0	244
Ireland	129.8	10.5	235
UK	111.2	12.2	233
Australia	106.3	11.7	223
Canada	94.7	12.2	217
Spain	119.1	9.6	215
Argentina	65.8	10.7	173
New Zealand	90.0	7.1	161
Mexico	40.7	2.1	62

Source: International Dairy Federation, www.fil-idf.org, with permission.

about 10 billion pounds of cheese per year. The states of Wisconsin and California are the largest producers, with an output that is more than three times that of the third largest producer, Idaho.

Europe, with a stock of 25 million lactating cows, roughly 2.5 times larger than that of the United States, is the largest producer of milk in the world. Overall EU dairy production is following the same trend toward larger, more specialized dairy farms, with an emphasis on obtaining the greatest amount of milk from each cow. The two breeds of cows that produce the greatest amount of milk—Holstein and Friesian—are now raised in most high-productivity dairy farms in the United States and Europe. Most dairy farms in the developed world use the CAFO mode of production, which is now becoming popular in Asian countries as well, including China, where the per capita consumption of milk has traditionally been very low.

ENVIRONMENTAL IMPACT OF DAIRY OPERATIONS

Although there are many similarities between the environmental impact of dairy cows and beef cattle, there are some important differences. A dairy cow is a prodigious producer of waste because of a much greater intake of food and water, a biological necessity for the production of large quantities of milk; it produces roughly 120 pounds of waste per day, about four to five times more than a beef cow and equivalent to that of 20 to 30 people. The total amount of wet manure produced by dairy cows in the United States is a staggering 464 billion pounds per year. Industrial dairy operations produce as much waste as midsized cities. But, unlike cities, which thoroughly process human waste in sewage treatment facilities, factory farms simply dump manure into uncovered manure lagoons and subsequently apply the waste to farmlands in neighboring areas. In some cases, this practice may endanger the health of local residents. For example, the *Cryptosporidium* contamination of Milwaukee's drinking water in 1993 killed more than 100 people and made 400,000 people sick.[3]

The feed of dairy cows, whether it consists of grains, seeds, or some other material, contains large amounts of nitrogen in the form of nitrates and phosphorus in the form of phosphates because these chemicals are essential components of nutrients in milk. About 30 percent of these feed nutrients are incorporated into the milk; the remainder is excreted in the manure that accumulates in the lagoons. Most high-productivity operations include fat, usually of animal origin, in the feed because the yield of milk increases by 3 to 8 percent for every pound of added fat.[4] Since heavy metals and toxins are stored in fats, feeding the fat of cattle and other animals to cows has the potential to contaminate milk with these chemicals, thus increasing the toxin content of milk and milk products.

Similar to other livestock operations, dairy farms emit noxious gases, including ammonia, hydrogen sulfide, methane, and VOCs, but the contribution of dairy cattle is greater on a per head basis. In North America, for example, the number of dairy cows is roughly one-tenth the number of beef cattle, but the amount of methane produced on dairy farms is about a third of that produced by feed-lot operations that raise beef cattle.[5] On the global level the ratio is not as skewed, because many dairy farms in developing countries are still operated in the traditional manner, but still the contribution of dairy cows is disproportionately greater than that of beef cows; less than a quarter of the cattle worldwide are dairy cows, but they are responsible for 34 percent of the total methane emitted from livestock operations. According to the EPA, dairy cows in the United States emitted two million metric tons of the greenhouse gas methane into the atmosphere in 2001. While most of this gas is released during the digestive process when the feed is fermented in cows' rumens, its emission during various stages of manure management is also substantial. Another greenhouse gas, nitrous oxide, is primarily released when ammonia from manure in the lagoons is broken down by microbes under anaerobic conditions. Compared to their numbers, dairy cows make a greater contribution to global warming than other kinds of cattle.

Annually, a dairy cow emits almost 20 pounds of VOCs and fine particulate matter—more than a car or a light truck.[6] VOCs are organic gases that react with the atmosphere to create smog and air pollution and are emitted both during the digestive process of the animals and from the waste generated by them. In the San Joaquin Valley in California, which has a large concentration of dairy farms with approximately 2.5 million cows total, dairies are the largest sources of VOCs and have an adverse impact on the quality of air in the valley. As mentioned earlier, the smallest particles are deposited

Table 8.3. Pollution produced by 13 million dairy cows and heifers in the United States, per year

Pollutant	*Amount (tons)*
Ammonia	705,435
Hydrogen sulfide	111,384
Volatile organic compounds	40,841
Particulate matter	22,277
Nitrous oxide	8,772
Methane	2,079,176

Source: U.S. Environmental Protection Agency, www.epa.gov.

in the lungs of people exposed to them, whereas larger particles are deposited in the airways of their respiratory tracts. Ammonia emissions are higher in intensively managed CAFOs than in older-style facilities. The operation of CAFO dairy farms depends on greater amounts of two precious resources—fossil fuels and water—than beef cattle feedlots of a comparable size, for several reasons. Housing dairy cows requires more frequent maintenance and more feed. These facilities also have to handle large volumes of milk for temporary storage. The energy requirements for transportation to a warehouse, packaging, and final delivery to the consumer are also much greater than for other foods of animal origin. Taking these factors into account, it is estimated that the production of each kcal of milk takes 16 kcal of energy in various forms.[7]

The drinking requirement of a dairy cow in the lactating phase is also very high and may be much as 30 gallons per day,[8] greater than the water requirement of a beef steer by almost a factor of three. Both the availability and quality of water are important to maintain the productivity of cows; limiting water availability would substantially lower the amount of milk produced. This factor is particularly important in parts of the world where water is not plentiful. The water given to dairy cows has to be free of contaminants such as high concentrations of minerals, high nitrogen content, bacterial contamination, pesticides or fertilizer products, and heavy growth of blue-green algae. The amount of water that a dairy cow needs on a daily basis increases rapidly with the temperature of the housing and the amount of dry matter in the feed. An increase in temperature from 60°F to 90°F increases the water requirement by about 30 percent.[9]

In addition to the water directly consumed by dairy cows, a much greater amount is needed for flushing the waste into holding tanks on a regular basis. In some facilities, the amount of water needed for flushing the waste and washing the animals may be as much as 150 gallons per day for each cow. Most of the water ingested by cows is returned in a polluted form containing animal waste, antibiotics, and hormones. Due to the large requirement of water for dairy facilities, the availability of water is the main consideration in deciding their locations.

The increasing demand for organic products, including milk, by consumers has resulted in more and more dairy farms dedicated to its production. A comprehensive assessment of the environmental impact of this change has to include, in addition to the dairy farms, the methods by which their feed is grown and the disposal of the waste produced by cows. Air pollution caused by the direct emission of particulate matter and VOCs from the dairy cows and the emission of ammonia and other gases from the accumulated waste will not be significantly reduced by changing from conventional to organic milk production. If synthetic fertilizers are not used on farmlands that grow the feed, the runoff of water will not be rich in nutrients, thus reducing its contamination both in the immediate neighborhood and in distant places. Since the feed of cows in organic dairy farms contains lesser amounts of cereal grains and more grass and hay, the emission of methane by cows, caused by fermentation of feed in the rumen, will be greater, thus causing a greater increase in global warming. However, this may be partly offset by reduced emissions of carbon dioxide and nitrous oxide gases. Organic milk production reduces pesticide use, but it increases land use per ton of milk. When all of these factors are taken into consideration, the adverse environmental impact of the production of organic milk is somewhat less than for conventional milk, but it is still substantial.[10]

HEALTH CONSIDERATIONS

Milk is a very nutritious food. It is an excellent source of calcium, phosphorus, riboflavin, and vitamin D and a good source of protein, vitamin A, potassium, and several B vitamins. An 8-ounce serving of milk contains eight grams of protein. The protein in milk is of high quality because it contains all of the essential amino acids in the proportion required by the human body. The amount of protein required by a person on a daily basis depends on his or her weight and level of physical activity. Proteins build and repair muscle tissues and serve as a source of energy during strenuous activities for extended periods of time.

Another important constituent of milk is calcium; an 8-ounce serving of milk contains 300 mg of calcium. The daily requirement of calcium increases from roughly 200 mg at birth to the peak value of about 1300 mg during the formative years and then slowly decreases to 1000 mg for mature adults. Calcium constitutes one-third of the mineral content of bones. It has to be consumed in adequate quantities on a regular basis to meet the demands for growth and to compensate for loss from the body in the form of sweat and other body fluids. Chronic dietary calcium deficiency eventually depletes bones and makes the person susceptible to osteoporosis and fractures. Dairy products have been positively associated with bone health largely, but not exclusively, because of their calcium content. A particularly persuasive study reported that low milk consumption during childhood was associated with doubling of hip fracture in a representative U.S. sample of postmenopausal women.[11]

The formation and maintenance of the integrity of bones requires the simultaneous presence of vitamin D and magnesium in addition to calcium. Vitamin D maintains the level of calcium in the blood within a narrow range to facilitate bone growth. It is also needed for proper functioning of the nervous system and helps the immune system in protecting the body from harmful invasions. Although milk is a rich source of vitamin D, this vitamin is destroyed in the pasteurization process and is added before releasing it to consumers. Human synthesis of this vitamin from sunlight decreases with age, thus increasing its daily requirement to some extent in our later years. Milk is also a good source of vitamin B12, which is rarely present in plant products. Vegetarian diets are generally deficient in this vitamin required for the production of red blood cells, which carry oxygen from the lungs to the muscles. An 8-ounce glass of milk provides 13 percent of the daily value of this vitamin.

A recent study has suggested that calcium is not the only thing in milk that is good for bones. Researchers led by Jilian Cornish of the University of Auckland, New Zealand, have found that an iron-binding protein, lactoferrin, present both in cow's milk and human breast milk, stimulates bone-forming cells in petri dishes and induces bone growth when injected into mice. While calcium provides the raw material for bone growth, lactoferrin and other unknown substances in milk appear to stimulate the activity of osteoblasts, the cells that form new bones. Lactoferrin also protects cells from apoptosis, a process that leads to the death of cells.[12] Milk has also been found to provide a certain amount of protection from colorectal cancer, reducing the risk of contracting that disease by about 15 percent. There are indications that some fatty acids present in milk may be instrumental in creating the benefits associated with dairy product consumption—bone health and reduced risk of stroke, metabolic syndrome, and some cancers.[13]

Although milk is a nutritious food, some concerns have been raised regarding its health benefits, including the possibility that it may even have some harmful effects, particularly when consumed in excessive amounts. Milk and milk products contain fats, including saturated fats, and cholesterol. Whole milk contains 3.5 percent total fats, 57 percent of which is of the saturated variety. Although low-fat products may have lesser amounts of fats, the fat content of cheese made from whole milk is very high. For example, typical preparations of cheddar, mozzarella, and Swiss cheese contain 33, 22, and 28 grams of fat, respectively, in each 100 grams, about two-thirds of which is of the saturated variety. Milk, cheese, and other dairy products, such as yogurt, butter, and ice cream, also contribute significant amounts of cholesterol to the diet. The established deleterious effect of saturated fats and cholesterol, particularly in relation to the risk of cardiovascular disease, provides one of the reasons for limiting the consumption of dairy products.

Although the evidence is only suggestive and not conclusive, some epidemiological studies have indicated that the consumption of excessive amounts of milk may have harmful effects that are not related to the presence of saturated fats and cholesterol. Most of these concerns spring from the hormones in milk. Even when a cow is not given synthetic growth hormone, its milk contains steroidal hormones such as estradiol, testosterone, and peptide hormones that include insulin growth factor, IGF-1.[14] Drinking milk has been shown to boost serum levels of IGF-1, which is identical in cows and humans. At this stage of knowledge, it is not possible to say with certainty if hormones in milk are bad for humans or if they have a protective effect. However, the presence of hormones in milk causes a certain amount of concern, even though most of them, including BST that is administered to some cows to increase the yield of milk, are present in the body anyway. The link between milk consumption and hormone-related cancers (breast and prostate) is debatable, with no general consensus.

Since milk is a good source of proteins that are easily assimilated by the body and also of other useful nutrients, its consumption is beneficial in parts of the world where the local diet in deficient in these substances. However, a typical Western diet contains more protein than recommended by health experts, and hence the consumption of milk, cheese, and yogurt has come under scrutiny and criticism. A few studies have indicated that excessive consumption of protein, particularly of animal origin, has a deleterious effect on the formation of bones because it causes them to lose minerals. An overload of protein may cause a condition known as calciuria that results in bone loss instead of bone formation and may also lead to the formation of kidney stones.[15] Since the typical diet in Western countries is already rich in proteins, an intake of large amounts of dairy products further adds to the load.

Due to the presence of saturated fats, cholesterol, and hormones in dairy products, it is often suggested that calcium be obtained from vegetable products. A number of items, such as green leafy vegetables, tofu, tapioca, almonds, and sesame seeds contain substantial amount of calcium, and many of them are also good sources of magnesium. Thus, they can be used to obtain desired amounts of calcium and magnesium without an overload of suspect chemicals. One problem with this approach is that milk appears to have other beneficial compounds besides calcium whose function is not properly understood at this time. While in principle it may be possible to get calcium and other nutrients from vegetable sources, it may not be generally achieved in practice. Information of this kind can be obtained by comparing the incidence of osteoporosis and fractures in populations that consume milk versus those that avoid it. Such studies have been conducted for both children and adults. In a study of children in New Zealand, the risk of fracture in those who avoided milk was 34.8 percent, while it was only 13.0 percent in those who drank milk on a regular basis. In a study of 11,619 Finnish women conducted from 1980 to 1989, those who did not drink milk because of lactose intolerance were found to have twice as many fractures as those who drank milk.[16] These results suggest that, although calcium and other micronutrients may be obtained from vegetable products, in practice it does not happen.

Vegetarians who include milk in their diet do not have compromised bone health; in fact, most epidemiological studies suggest a positive correlation between the intake of milk and bone density. Vegans who exclude dairy products from their diets have a higher fracture risk, according to data obtained by the European Prospective Investigation into Cancer and Nutrition.[17] The nutrients a person is most likely to be deficient in if milk products are excluded are calcium, potassium, and magnesium. Important evidence that supported including 3 cups of milk or milk products daily came from the CARDIA (Coronary Artery Risk Development in Young Adults) study, a 10-year longitudinal study of 3,157 black and white adults aged 18-30 years from 4 U.S. cities. Each daily serving of dairy lowered the risk of developing insulin resistance syndrome, which is characterized by obesity, hyperinsulinemia, insulin resistance, and hypertension, by 21 percent. Best results were achieved when about three servings of dairy products were consumed daily.[18]

Milk from all sources contains substantial amounts of a compound known as lactose. Human infants produce the enzyme lactase, which allows them to digest human milk and can also help with the digestion of cows' milk. While some people keep producing this enzyme in later years, in others its biosynthesis declines sharply after the age of five. An inability to produce lactase makes them lactose intolerant, so that they cannot digest milk or milk products; drinking milk causes bloating and abdominal discomfort. However, the production of lactase is not completely shut off, and most people can consume products that contain reduced amounts of lactose, such as yogurt and some cheeses. The lactose content of cheeses, in general, is much lower than that of milk; hard cheeses that are made by extended periods of fermentation have insignificant amount of lactose and hence can be consumed by most people.

* * *

Although dairy products appear to have numerous advantages for health and well-being, it is perhaps safe to advise that the consumption of milk and its products be limited. Organic milk may not contain pesticides and antibiotics that nonorganic milk contains but still can be loaded with fat and cholesterol. Even organic cows' milk, which does not contain artificial hormones, contains naturally occurring hormones. In this connection, it is significant that pasteurization reduces protein hormones (IGF-1) in milk by 10 to 15 percent, and ultrahigh-temperature processing, which makes it possible to keep milk at room temperature and is slowly becoming popular, removes even more. Environmental and health considerations call for a substantial reduction in the consumption of milk and dairy products from the present high levels in developed countries. Consuming lesser amounts of milk, yogurt, and cheeses will take pressure away from the dairy industry to produce excessive amounts of milk by feeding grains that could be directly consumed by people to cows instead. A large reduction in the demand for milk would be beneficial both for the environment and for human health.

9 * SUFFERING

Animals raised for meat will eventually be killed to get access to their flesh, dairy cows will be milked, and eggs will be collected from laying hens. None of these circumstances means that animals have to lead a painful existence and endure suffering throughout their lives—suffering that is obvious to even a casual observer and can be proved by a pathologist's examination of their carcasses.

Traditional animal farms saw an intimate bond between the animals and their keepers. Although the ultimate fate of the animals was sealed, the owner of a small farm had to be concerned about the health and welfare of his stock, partly because keeping the animals healthy was good for his business. The farmer also developed a certain amount of empathy with the animals because he or she observed the animals closely every day.

The transition from family farms to Concentrated Animal Feeding Operations is similar to the change that has taken place in the production of objects such as furniture: rather than craftsmen creating an object in a workshop, we have assembly-line production in factories. While a craftsman can look at her or his work with pride and satisfaction, a factory worker repetitively performs the same task involving just one component of the product, with which he or she has no real connection. In the industrial mode of food production, an individual animal loses importance; the sole emphasis is on the bottom line, determined only by the value of total inputs and outputs for these operations. A herd of replacement animals stand ready at all times to take the place of any animals that are not performing at the optimum level, and as a consequence, each individual animal is considerably less important.

As we have seen, animals in CAFOs live in environments that are not compatible with their nature and that are detrimental to their health—they are fed an unnatural, energy-rich diet and packed tightly in a space that is the bare minimum for their survival. In these conditions, they would quickly succumb to infections or disease were it not for the antibiotics and vaccines that are administered to them. It also benefits the CAFOs that the animals' lives are mercifully short—they live only while they are rapidly putting on weight. In human terms, they are slaughtered at the end of their teenage years. Animals raised in industrial facilities are tortured: their bodies are mutilated, and they suffer psychological deprivations, physical problems caused by unnatural feed, and genetic manipulations performed with the sole objective of increasing the output of the desired food item. Forensic examination of their bodies after slaughter provides evidence of suffering caused by broken bones, diseased organs, cuts and bruises, and so on, and their psychological suffering can be inferred from an observation of their behavior pattern.

A consideration of physical and psychological suffering assumes that animals have the capacity to feel pain, that they are capable of emotions such as fear and distress, that they have memory of past events and expectations for the future. In the seventeenth century some philosophers declared that animals were machines with no mind or feelings. Animals, these philosophers declared, do not really experience pain; their reaction to injuries is simply a mechanical response. And, since they were considered to be machines with no capacity to feel or think, how they were kept was completely unimportant. This position is at least tacitly accepted by many omnivores. One large contradiction stands in the way of this thinking, however: we know that society accepts that pets of all kinds, including birds, fish, and reptiles, have a capacity to feel pain, because society passes laws against cruelty to these animals.

The view that animals are like machines that live only in the present has now been soundly discredited by scientific research. While all animals feel simple sensations like hunger, pain and fear, those that are used for food are social animals with a more advanced level of sentience involving recognition, memory, understanding of the present, and expectations for the future. Cows, sheep, chickens, and pigs are sociable, intelligent creatures that bond with their keepers and also among themselves. Contrary to the usual perception, chickens are not mindless, simple automata but are complex behaviorally, are capable of learning, show rich social organization, and have a diverse repertoire of calls.

When cows are kept in a fenced area, they explore the entire region to determine the limits of their enclosure; an analysis of their brainwaves shows that cows get excited when they succeed in solving a puzzle. Numerous studies have shown that birds may plan for breakfast for the next day,[1] chickens prefer food laced with drugs that will relieve their pain over food without the painkilling drugs,[2] and pigs are as intelligent as dogs. It has been conclusively shown that animals raised for food are endowed with a good measure of intelligence and are social creatures with a capacity to remember humans and others for long periods. The important question, however, is not whether these animals can reason, remember, and have the mental acumen to solve puzzles—faculties that they are known to possess—but whether physical hardships and psychological deprivations in industrial operations cause them to suffer from physical pain and mental anguish for substantial portions of their lives.

All living creatures must be able to feel pain in order to survive. When a pharmaceutical company develops a new analgesic to relieve pain in humans, it is first tested on animals such as mice and hamsters (animals that show less intelligence than farm animals) for efficacy; this animal testing is further evidence of the acceptance of the fact that animals feel pain. In an experiment, chickens that suffered from neuromuscular pain caused by lameness were given a choice of foods, one of which was laced with the analgesic carprofen. Researchers found that these chickens preferred to take the feed laced with the drug, thereby proving that lameness was causing chronic pain, which the birds preferred to avoid by taking the drug, just as humans would in similar circumstances.[3]

Five aspects of the lives of farm animals in CAFO settings that give them intense discomfort and pain are:

1. extreme confinement and filthy living quarters
2. husbandry practices of mutilating their bodies to prevent problems caused by unnatural living conditions
3. problems caused by manipulation of the physiology of animals using genetic engineering, high-energy feed, and growth hormones
4. transport of animals over long distances
5. methods used for slaughtering

Other factors aggravate the pain and discomfort. Since farm animals are social animals, an inability to form bonds with their kind causes additional anguish. In confined operations, these creatures are not even able to do the minimal things that come naturally to them—chickens spreading their wings or pecking for grains or insects, or mothers tending to their young.

Farm animals have been selectively bred for centuries, but new technology has accelerated the process and, through genetic engineering, greatly increased the level of manipulation that can be achieved. Their feed is carefully formulated to increase the desired output—meat, milk, or eggs—and even includes items that are contrary to the nature of the animals, such as seafood and fats of other animals. Unusable meat and animal fat, even waste products in the case of chickens, are recycled over and over again to make use of every bit of energy. The result of these innovations, as discussed in chapter 1, is that animals become disproportionately large in a very short time—a broiler chicken reaches the desired weight of four pounds in only six weeks, and steer are sent to the slaughterhouse before they are two years old.

PHYSICAL EVIDENCE OF PAIN AND SUFFERING

Chicken CAFOs raise 200,000 to 300,000 birds at any given time, with birds at different stages of growth kept in separate buildings for a continuous supply of meat. Each building houses 20,000 to

30,000 birds and is bare, except for dispensers providing food and water. In the beginning, they have some room to move around, but the space decreases as the birds grow bigger, so that eventually they are packed tightly in their enclosures. In these cramped conditions, a bird cannot even open its wings. The housing is filthy, with piles of excrement that harbor various types of pathogens, including *Salmonella,* which is nearly ubiquitous in factory-raised chicken. Until workers remove them, any birds that die remain with the living birds. The flock is taken out of the warehouses only at the time of slaughter.

Genetic engineering has been used to produce chickens with large breasts and thighs, since these are the most desired parts of a carcass. The rapid growth of their bodies does not allow enough time for their bones to develop in corresponding proportions, frequently resulting in painful skeletal deformities and inflammation of joints, since their legs cannot support their abnormally heavy bodies. Up to 30 percent of the birds (chickens, ducks, and turkeys) have dyschrondroplasia, an abnormal growth of cartilage and swelling of joints caused by weak, bowed legs. In extreme cases, this may even result in broken bones,[4] forcing the birds to lead a painful existence until the end of their lives.

The hearts and lungs of these birds are also not developed enough to support their massive bodies, resulting in the deaths of millions of birds before they reach market weight. In addition to these problems, high stocking density increases the incidence of blisters in breasts, chronic dermatitis, hock burns, and various other infections. Turkeys raised in the United States are now so obese that they cannot reproduce naturally; instead, all turkeys born in this country are conceived through artificial insemination.[5] Improper temperature and humidity, dust particles in the air, and ammonia gas produced by accumulated filth combine to create an unhealthy and painful environment for the broilers.

Laying hens are kept in battery cages made of wire mesh, usually eight or nine birds in a cage. These cages are built one upon another in sheds that hold tens of thousands of birds. Each hen has an average of 67 square inches of space—less than the size of a standard sheet of paper; thus a hen cannot stand up straight or flap her wings throughout her life. This is very stressful because, under normal conditions, birds find a quiet place to sit while laying eggs. Constant egg production as part of this miserable existence, without being able to engage in any normal activity, leaches calcium from their bones, causing severe osteoporosis and broken bones. Additionally, the wires of the cage injure their feet, as the hens must sit in essentially one position for their whole lives with their feet pressing into the wires. Sloping floors of battery cages that facilitate collection of eggs injure the hens' feet as they continually try to maintain balance while they slip on the inclined surfaces. Any attempt by the birds to move in these cramped conditions results in injuries and damages their feathers.

The most important characteristic of a breed used for laying is that it should produce the greatest number of eggs possible per year. Manipulations of genetics, environment, and feed during the last few decades have forced hens to produce more than double the number of eggs, so that they now produce eggs, on average, nine days out of ten. In natural settings, hens go through a period of molting, usually induced by reduced daylight in the fall. At that time ovulation and the production of eggs stop almost completely. The diminution of sex hormones leads to new feather growth, which forces out old feathers. However, waiting for the cycle to proceed naturally is not cost-effective because of the diminished quantity of eggs produced during that period. With the artificial light and controlled environment of industrial operations, the quantity of eggs is independent of seasons and declines after about one year. Researchers discovered a few decades ago that "forced molting" can be induced in hens in a very short time by keeping them in conditions of extreme deprivation and stress, so that ovulation ceases. Factories that use forced molting deprive the hens of food and water for one or two weeks and keep them in the dark. They typically lose 30 to 35 percent of their body weight during this period due to thirst and starvation, and quite a few die as a result of these deprivations; however, a sufficient number survive to make this process economically useful. When food is given to the hens after the molt, their egg production resumes, although somewhat below the previous level. This method is attractive because hens start producing eggs again after a gap of only two weeks. After two or three such cycles, the hens are completely worn out and are slaughtered. The flesh of laying hens is not good enough to be sold as such and is turned into chicken stock or low-grade meats. Hens that are not force-molted are slaughtered after the first year. Forced molting has been banned in some European countries, and there are movements afoot to ban it in others.

As a breed, laying hens are not suitable for the production of meat because they will not develop large muscles in the thighs and breast and will not reach slaughter weight as quickly as desired. For this reason, the male chicks of laying hens are ground up or suffocated as soon as their sex is identified. Popular magazines show photographs of newly hatched chicks as the most sensitive and delicate creatures; about 200 million male chicks are crushed to death each year in the United States. These deaths are an outcome of advances in genetic engineering that produce species that are highly specialized for a single purpose and hence useless for any other.

Breeding sows, used as factories for producing piglets, lead an even more miserable and painful existence than chickens. After being artificially impregnated, they are kept in metal crates that are almost the same size as their bodies. As described earlier, the space in these gestation crates is so tight that the sows can only rest on their hind legs and are unable to turn around. Similar to the cages for breeding hens, the gestation crates are usually piled on top of each other so that the filth from upper cages drips onto the animals in lower cages. The economic advantage of this extreme confinement is that it allows a great many sows to be housed in an environmentally controlled space with a minimal consumption of energy and allows a small staff to take care of their basic needs.

Extreme confinement in these highly restrictive crates puts serious strain on the physical and mental well-being of the sows. The unnatural flooring and lack of exercise causes obesity and crippling leg disorders. Sows often suffer from infected and swollen teats and joints. Extreme confinement makes their skin tender and susceptible to infection, particularly when the sows rub against the bars in these dirty conditions. Sows often suffer from long-term pain caused by infected cuts and abrasions, foot injuries, and lameness. Lack of exercise leads to weakened bones and muscles, heart problems, and urinary infections. At the end of their four months of pregnancy, they are transferred to similarly cramped farrowing crates to give birth.

Sows typically give birth to nine or ten piglets per litter. The industry seeks to minimize the time between pregnancies by weaning the piglets only a week after birth, although in some cases they are allowed to stay with their mothers a bit longer. The sows are then artificially inseminated and the process begins all over again, with the result that each sow has more than 20 piglets per year. The continuous cycle of pregnancy and giving birth wears out the body of the sows in a short time, and so a typical sow is slaughtered at about three years of age. Against all their instincts, sows must give birth to piglets, nurse them, eat, sleep, and defecate, all in the same cramped space that does not even allow them to turn around.

Piglets are moved to "finishing" facilities—large, barren chambers—when they are about two months old and are kept there until they are ready for slaughter. Although pigs in natural surroundings make nests of straw, the floors of these pigs' stalls are kept bare to make it easy to flush out manure and waste material. In these facilities, about 100 pigs are kept in a metal pen with slatted concrete floors. The total number of pigs being raised at one time in a facility can easily exceed ten thousand. Just like other animals, pigs are routinely administered antibiotics, vitamins, and hormones to prevent the spread of disease and to facilitate their bodily growth.

Cattle also lead a painful existence. The living quarters of cattle in feedlots are invariably filthy. Since the packing density of the animals is extremely high, it would be prohibitively expensive or impossible to remove waste very often and, as a result, most animals have to live in accumulated excrement, contrary to their habits in natural surroundings. As a consequence of living in filth, standing on hard surfaces, and having no activity, many cattle become lame and lead a life of extreme discomfort and pain. An audit in 1999 found that 26 percent of beef cows, 39 percent of dairy cows, 36 percent of beef bulls, and 29 percent of dairy bulls had deformities in their legs and that the proportion of cattle suffering from lameness was greater in that year than in 1994.[6]

Lameness and leg injuries are most common in dairy cows, partly because their period of confinement is greater and their stalls usually have hard surfaces. They are made to produce enormous amount of milk, and the volume of milk they produce causes inflammation of the sensitive cells in their udders and results in a painful condition known as mastitis. This condition is further aggravated by dirty living conditions, since the cows often have to lie in their own excrement.

In natural conditions, cows use their long tails to remove flies and insects from their bodies. But in confined operations, where the stalls accumulate filth, the tails spread feces on workers in the stalls, and so, to prevent this, the tails of dairy cows are often cut off a few weeks before the first calf is born. The industry's argument for "tail docking" is that it improves cleanliness because it prevents the tail from transferring feces to the cow. However, a study of 500 dairy cows, half of whom had their tails cut off, found no difference between cows with intact tails and those that had their tails docked in terms of any of the cleanliness measures, somatic cell counts (a measure of udder health), or cases of mastitis.[7] These results suggest that this procedure, while improving the comfort of workers, substantially adds to the pain and suffering of the cows. Docking the tails of cows is a painful procedure in itself and makes the cows miserable because they are not able to use their tails to remove flies or insects that may be biting them. A number of studies have found more flies and insects on tail-docked animals. Several European countries, including Norway, Sweden, the Netherlands, the United Kingdom, and Switzerland, have prohibited tail docking of dairy cattle. However, at present there is no legislation in the United States to address this issue.

The demand for quick turnover requires that cattle be kept on an aggressive grain-feeding program instead of eating hay and forage. Their feed also contains fats that have been generated in other phases of industrial operations. This unnatural diet causes acute acidosis and is dangerous for the proper functioning of the liver, which becomes infected with harmful microbes, making the animals very sick. Despite administering drugs to prevent acidosis and bacterial infection, in some cases the liver becomes so enlarged that a steer cannot eat anything and has to be culled. An examination of the carcasses of slaughtered cattle showed that up to 32 percent suffered from liver abscesses,[8] indicating a painful existence for a long period. Routine examination of the carcasses of slaughtered animals found that many other organs also show signs of decay and have to be discarded, even though the animals are slaughtered at a very young age. In addition to the liver, which is the main site that is affected by the ingestion of unnatural feed, these organs include the stomach, heart, and lungs. Other evidence of suffering is provided by the condition of the hide after the animal is slaughtered: damage to the skin indicates that the animal was subjected to painful situations for extended periods. The most prominent damage to the hide is caused by hot-iron branding and attacks by insects while the animal was alive. Because of such damage, a substantial fraction of hides is discarded.

Beef cattle are given so many intramuscular injections that they form lesions in the top sirloin butt, and a substantial part of this prized cut of meat must be discarded. The amount of degraded meat is large enough to become a cause of concern for the livestock industry.[9] It is obvious that injuries that degrade the quality of muscles and make them inedible must be a source of constant pain for the animals.

Another source of pain is the removal of horns and testicles. The horns of all cattle are destoyed early on to prevent injury to workers and to other animals in the herd. There are three different methods for removing the horn buds when the calf is young, each of them very painful for the animal. In the chemical method, caustic potash or caustic soda is applied to the horn bud and is allowed to burn the area in a matter of days or weeks. The hot-iron method uses a specially designed iron to destroy the horn button tissues at the base of the bud, and, in the commonly used mechanical method, a tube with a cutting edge is attached to the base of the bud. It is pushed and twisted until the newly formed bone has been entirely scooped out. Anesthetic is seldom used with any of these procedures. Also without the use of sedatives or anesthetics, male beef cattle are castrated, because the meat of a steer commands a higher price than that of a bull.

INDICATIONS OF STRESS

While a forensic examination of a carcass can provide evidence of physical injury that, we can infer, must have caused a painful existence, aberrant behavior points to the amount of psychological stress on the animals caused by unnatural living conditions, treatment, and handling. Such conditions also cause stress hormones to be released into the bloodstream, and these hormones adversely affect the quality of meat of the slaughtered animals. Animals exhibit many signs of stress:

Piglets make distinctive, frequent squeals when they are separated from their mother, and some of them eventually become unresponsive.

The heart rates of sheep increase when they cannot see other sheep.

Pigs, calves, and cows that have been mistreated try to avoid humans.

Pigs can collapse and die as a result of rough handling by humans.

Cows express obvious signs of distress when they are separated from their calves.

In addition to signs of suffering and extreme discomfort, these animals often develop behavior patterns that are contrary to their nature and survival instinct, indicating that the psychological stress causes them to lose their mental balance. While in natural surroundings, all females protect their young, even at the risk of their own lives; in high-yield factories, laying hens sometime start crushing their eggs and eating the contents, something that never happens in nature. Chickens also peck each other, causing serious injury and even death. Pigs and steer bite everything in sight, including the bars of their cages; this breaks their teeth, reduces their ability to feed, and even causes premature death. A pig may bite the tail of another pig, causing grievous injury. The victim of tail biting gradually ceases to react to being bitten, exhibiting what we would call learned helplessness in humans. Such behavior patterns are indicative of extreme distress in the animals caused by harmful living conditions, extreme confinement, and lack of opportunity to socialize with other animals.

The livestock industry's solution to these aberrant behaviors is additional mutilation, which causes even more suffering. To guard against the possibility of chickens' injuring each other, the standard practice is to cut off a large section of their beaks. This so-called debeaking is performed with a hot blade. The severed nerve endings in the beak can develop abnormal nerve tissue, and the beak never heals properly. For many years, the industry argued that beak trimming was a benign procedure, analogous to cutting nails in humans. However, it is now clear that this is not the case and that the trimming causes behavioral and neurophysiological changes. According to Michael Gentle and his colleagues at the Institute of Animal Physiology in Edinburgh, debeaked chickens suffer from acute and chronic pain following the operation.[10] After hot-blade trimming, damaged nerves develop into extensive neuromas that slowly begin to discharge fluids and are known to be painful in humans and animals. Chickens in natural conditions continuously peck in search of food, but debeaked chickens in cramped surroundings are able to use their beaks to eat food only from dispensers. In some cases, debeaked chickens die because they are not able to eat at all. The toes as well as the beaks of turkeys are often cut off with a hot blade.

Pigs in the wild are highly social and clean creatures, naturally friendly, loyal, and forgiving. They are also quite intelligent, at least as intelligent as dogs, and can be taught to play games and other activities. Piglets learn their names by two to three weeks of age and respond when called. They are extremely active, inquisitive, and careful to maintain their living space. They graze and use their keen sense of smell and sensitive snout to find buried roots, shoots, worms, and larvae. They are social animals; at night, up to 15 animals will snuggle together to keep warm. Expectant sows build nests and line them with leaves; they are attentive to the 7–14 piglets born to them. Since they do not have sweat glands, pigs cool themselves by wading in water or wallowing in mud, if that is all that is available to them. If they are given enough space, pigs are careful not to soil the areas where they sleep or eat. But in factory farms, they are forced to live in their own feces, urine, and vomit and even amid the corpses of other pigs. Overcrowding and confinement in large production operations are unnatural and produce stress.

The conditions in the stalls in which the pigs are raised are completely opposite from the conditions in their natural setting; these stalls are completely unsuited to the pigs' physical and behavioral needs. Tightly confined, with no space to teach their piglets normal behaviors essential to their development, sows frequently develop neurotic behaviors, such as gnawing on the bars of their crates, injuring their teeth and making eating difficult or impossible. Pigs in natural settings do not attack or injure each other, so it is believed that the aberrant behavior of tail biting is the result of stress caused by living in the overcrowded environment of factory farms where pigs are deprived of all normal stimulation and activities. To prevent tail biting, their tails are cut off at an early stage, and sometimes a few teeth are also removed. Male pigs are subjected to painful castration to diminish aggression and to prevent the production of adult male sex hormones. Such surgical interventions, which are almost always done without the use of anesthetic, remove some undesirable behaviors, but they aggravate the suffering of these creatures.

Even the end of life for these animals is not simple or direct and causes enormous suffering. Environmental regulations, community concerns, and other prohibitions have greatly decreased the number of slaughterhouses and made them high-volume operations, generally located far from the feedlots where the animals are raised. Livestock are usually transported over long distances with insufficient food, water, and temperature control—transport that causes extreme distress and, in many cases, physical injury. According to industry reports, millions of animals die or become crippled during transportation.

Converting a 1,200-pound steer into pieces of meat is never going to be pretty, but the actual process in many of these operations represents the height of cruelty. Steers are forced to move in single file by giving them electric shocks and are then transferred to a conveyer belt. One person fires an eight-inch lead bolt into the heads of the animals as the conveyer belt transports them at the rate of about five per minute. This pace does not allow time to find out whether the animal has died, as intended, while it is moved to sections of the factory where the dismemberment of bodies begins—cutting off the hooves, skinning, and so on. The result is that the animals die "piece by piece."[11]

Industrial animal operations have the sole objective of maximum productivity at minimum cost. Workers have little stake in the totality of the operation, and they perform repetitive tasks. Some of them commit acts of terrible cruelty to the animals. Work in the slaughterhouses is so low paid, dangerous, and demeaning that the industry often has difficulty finding workers to fill the positions.[12] The injury rate among these workers is among the highest of any occupation in the United States.

The slaughter of chickens is completely automated in large facilities; they are dipped in electrified water to render them insensible before their throats are cut and their bodies moved to a scalding bath to remove their feathers, with no certainty that the birds are completely stunned. The USDA oversees the treatment of animals in these plants, but the enforcement of the law is difficult, and fines are rarely imposed. Robert Byrd, the longest-serving member in U.S. Senate history, said in a speech on the Senate floor in 2001, "Our inhumane treatment of livestock is becoming widespread and more and more barbaric . . . such treatment of helpless, defenseless creatures must not be tolerated even if these animals are being raised for food."[13]

* * *

Most of us do not know—or choose not to know—how meat is made, because intensive production systems, completely shielded from the general public, allow us the luxury of not thinking about the implications of factory farming. In a very short period, raising livestock has morphed into an industrial endeavor that bears little relation to the landscape or the natural tendencies of the animals.

In general, we develop concern and compassion for all creatures around us, except for those who pose a threat to us or to our belongings. Cat and dog owners are concerned about their pets, and the owner of a hamster, a rabbit, a reptile, or a bird observes the animal daily and develops a bond with the creature. When a racehorse breaks its leg and is euthanized, the story makes headlines, because people are genuinely concerned about its suffering.[14] Many people who eat animal products would find the cruelties committed against animals being raised for food highly repugnant. This is why the industry must build a thick wall of secrecy around these operations and why consumers pick up a plastic-wrapped package of meat in the supermarket the same way they pick up a bag of potatoes—without giving a thought to the history of that package.

The knowledge that our taste for animal flesh is causing immense suffering to animals throughout their lives may force many of us to think twice before eating a meal. An awareness of how and where food is raised will inevitably have an impact on the amount of animal food that is produced.

10 * CONSEQUENCES

Previous chapters documented what happens to the quality of land, air, and water—and humans—when we eat foods of animal origin. Producing animals for consumption is degrading the environment and adversely affecting the productivity of farmlands at a time when the demand for animal products is increasing. Raising animals in industrial, warehouse-type facilities harms the environment more than other methods of raising them do and creates new problems, such as of the growth of dangerous pathogens. The extensive list of adverse health effects of eating animal-based foods and the adverse environmental effects of the common methods of raising farm animals raise questions. What about alternative dietary patterns? What are the reasons for the growth of the CAFO mode of production? What are the possibilities for raising livestock in a way that is less harmful to the environment? A related question is this: Can science solve the problems of increasing agricultural productivity so that the present modes of consumption and production can be sustained without making major changes in our diet or our methods of raising livestock?

VEGETARIANISM THROUGH THE AGES

Vegetarianism has never caught the imagination of the Western world, where it has remained a fringe movement, occasionally advocated by a few notable figures. The earliest known Western philosopher who was against the killing of animals for food was Pythagoras (570–490 BCE), some of whose ideas were saved for posterity by his followers, particularly Ovid. Pythagoras believed that the soul is immortal and transmigrates into other kinds of animals, thus establishing a fellowship between all living things. Ovid wrote of Pythagoras' revulsion toward killing animals for food in this passage from his *Metamorphoses*:

> Oh, what a wicked thing it is for flesh
> To be the tomb of flesh, for the body's craving
> To fatten on the body of another,
> For one live creature to continue living
> Through one live creature's death. In all the richness
> That Earth, the best of mothers, tenders to us,
> Does nothing please except to chew and mangle
> The flesh of slaughtered animals?[1]

Other scholars in the ancient world who supported vegetarianism, such as Seneca (4 BCE–65 CE) and Plutarch (56–120 CE), believed that abstaining from meat purifies the spirit and hence elevates an individual to a higher level of knowledge and understanding. For the philosopher Porphyry (234–305 CE), an individual's compassion for animals, on the one hand, and the adverse effects of ingesting the flesh of animals, on the other, were intimately connected to emotional and intellectual development.

A number of Christian thinkers, beginning with Saint Benedict in the sixth century, defended a largely plant-based diet; their advocacy was primarily based on religious rather than ethical reasons, and their influence was mostly limited to monastic circles. Vegetarianism was not considered essential or desirable even by those who had forsaken worldly possessions. Saint Francis of Assisi preached that all creatures, from the highest to the lowest, are equivalent in the eyes of God, and that nature is not to be dominated, but he did not prohibit his followers from eating meat except during times of fasting—and for ascetic reasons rather than out of concern for animals.[2]

Influential advocates of vegetarianism in the nineteenth century included the poet Percy Bysshe Shelley (1792–1822) and the writer Leo Tolstoy (1828–1910). Shelley saw vegetarianism as a necessary step in the moral perfection of humanity and believed that eating the flesh of animals leads to all the baser instincts in men:

> He slays the lamb that looks him in the face,
> And horribly devours his mangled flesh,
> Which, still avenging Nature's broken law,
> Kindled all putrid humours in his frame,
> All evil passions, and all vain belief,
> Hatred, despair, and loathing in his mind,
> The germs of misery, death, disease, and crime.[3]

Tolstoy adopted a vegetarian diet in 1855 after seeing how inhumanely and callously cattle were butchered. He described in graphic details the horrible scenes he witnessed in an abattoir. For him, eating the flesh of animals was immoral because it involved a brutal act.

Notable champions of vegetarianism in the early part of the twentieth century include Romain Rolland (1866–1944), Albert Schweitzer (1875–1965), and Mohandas Gandhi (1869–1948). More recently, Tom Regan, a professor of philosophy at North Carolina State University who publishes widely on animal rights, wrote an article titled "Utilitarianism, Vegetarianism, and Animal Rights." He advocated that the concept of moral worth, reserved by Immanuel Kant for human beings, should be extended to animals as well.[4] The Princeton philosopher and bioethicist Peter Singer argued in the book *Animal Liberation* that there can be no moral justification for refusing to take the pain and suffering of animals into consideration.[5]

The vegetarian movement in North America has generally stressed the health-related benefits of a diet rich in fruits, vegetables, and grains. Seventh-day Adventists, who established their headquarters in Battle Creek, Michigan, in 1855, also emphasized a vegetarian diet for its beneficial effects on health. To some extent, they have helped in spreading vegetarian principles and educating people about the benefits of a diet of plant products. Although their religion does not mandate vegetarianism, today about half of the estimated five hundred thousand Seventh-Day Adventists in the United States follow a vegetarian diet. A number of epidemiological studies on the effects of diet on health have included vegetarian Seventh-Day Adventists for comparison with nonvegetarian Americans and have generally established the benefits of the vegetarian diet. Nevertheless, the general public has not enthusiastically embraced vegetarianism.

Surveys seeking to calculate the number of vegetarians in the country run into conflicting definitions of vegetarianism. People who include fish in their diet, for example, still declare themselves to be vegetarians, as do many people who eat meat on an occasional basis. The 2008 "Vegetarianism in America" study, published by *Vegetarian Times* magazine, puts the number of U.S. adult vegetarians at 7.3 million, or 3.2 percent of the noninstitutionalized population. The percentage of vegetarians in other developed countries is not significantly different from this number. The study found that female vegetarians outnumber males by almost two to one.

Many people associate vegetarianism with weight-loss programs. This false connection is unfortunate, because it induces some people to engage in unhealthy eating styles. Recently Ramona Robinson-O'Brien, a professor of nutrition at the College of Saint Benedict–Saint John's University, and others studied 2,516 teenagers and young adults to identify their eating habits, vegetarian status, dietary quality, weight, physical activity, binge-eating practices, healthy and unhealthy weight-control behaviors, and substance abuse. They found that only 4 percent of the participants in this group were vegetarians, while 11 percent stated that they used to be vegetarians. In this study, vegetarianism was associated with a number of benefits, including a lower BMI, a lower rate of obesity, greater consumption of fruits and vegetables, and lower consumption of calories from fats. The study also found, however, that teen vegetarians were twice as likely to engage in unhealthy weight-loss behavior.[6]

Historically, India has been the home of most vegetarian traditions since the earliest recorded times. Part of the reason can be traced to the interconnected religions that grew out of that land—

Hinduism, Buddhism, and Jainism—each of which gives special reverence to all forms of life. An ancient Hindu book that has been venerated through the ages is the *Bhagavadgita*. It divides foods into three classes: *saatvik, raajas,* and *taamas*. Although the book does not describe the contents of these diets in great detail, it is widely understood that saatvik food consists of fruits, vegetables, whole grains, nuts, and items that have not been heavily processed. The book says that this kind of a diet promotes longevity, intelligence, vigor, health, happiness, and cheerfulness. The raajas diet, consisting of fried, heavily spiced, and processed foods, is inferior to the saatvik diet. Finally, the taamas diet, consisting of animal-based foods, promotes qualities opposite to those promoted by saatvik foods.

Jainism and Buddhism were stricter than Hinduism in demanding adherence to a vegetarian diet. Due to the interaction of these religions, which flowered in India centuries before the Common Era, India became almost entirely vegetarian, and this tradition became deeply rooted. A Chinese Buddhist monk, Fa Hsien, went to India toward the end of the fourth century CE in search of authentic Buddhist books. He spent a few years in India, traveling from the North to the South, and left the country in the early years of the fifth century CE. His travelogue declares India to be a strange country where people do not kill any living creatures, do not keep pigs or fowl, and do not sell live cattle.[7]

Indian vegetarianism has been widely influential. Taoism, one of two ancient religions of China (the other is Buddhism), is heavily influenced by Hindu yogic philosophy. The first Tao book, *Tao Te Ching,* written by the Chinese sage Lao Tzu, is essentially a manual on yoga. From the earliest times, meditation and vegetarianism have been considered essential components of yoga. Like Buddhism, Taoism requires a vegetarian diet. However, just as with Buddhism, the practice of vegetarianism is limited to the strongest adherents to the faith.

To consider another religious tradition, the practice of Islam involves eating meat almost as a prerequisite, but during centuries of rule over India, some Muslim groups developed Sufi sects that practice vegetarianism, and they have also acquired some other Hindu practices. There are Sufi sects in many Islamic countries in Asia and Africa.

Centuries of rule by Muslims and Western powers has slowly diminished the traditional vegetarianism of Indian society. Just as with people living in much of the developing world, there is a desire among most Indians to emulate the lifestyle in Western countries. The result is that the number of committed vegetarians has precipitously declined in recent years. One report indicates that almost two-thirds of the Indian population may eat meat whenever possible.[8] Eating meat is particularly popular among the emerging middle class in that country. However, due to social and financial constraints, the per capita consumption of meat in India is the lowest in the world.

Vegetarianism was considered to be strange and mysterious in the Western world until a few decades ago. However, the attitude of the general public has gradually changed. Results of scientific studies that establish the health benefits of a diet rich in fruit, vegetables, and whole grains are publicized in the media, and the number of persons who accept the benefits of a vegetarian diet may be gradually increasing, although at a very slow rate. For a variety of reasons, many people who accept the health benefits of a vegetarian diet do not necessarily follow it.

MODERN LIVESTOCK OPERATIONS

The main reason for the growth of the modern industrial mode of raising livestock is that it allows a few corporations to control this huge and highly lucrative industry. Advocates of CAFOs can also argue that these operations have made animal products abundant at reasonable prices. Although a number of scientific innovations have helped in the growth of CAFOs, the most important reason for the low price of their products is that the cost of environmental degradation, health hazards, and use of nonrenewable resources are not taken into consideration. These costs must be handled by the communities where the CAFOs are located and, truly, by the entire world. If these aspects are taken into account, it is clear that far from being cheap, the real price of the current food production system is enormous: it is imposing a staggering monetary burden on us and even more on future generations, because of the cumulative effect of these factors.

As we have seen, there are other factors that contribute to the low price of meat for the consumer:

- Animal parts such as fats, blood, seafood bycatch, and even feces in some cases, are recycled by feeding them to other animals after sanitization. This recycling procedure makes maximum use of the available energy, but it increases the chances of contamination of the food.
- Automation of all phases of operation allows a small staff to take care of a large number of animals, thereby reducing the cost of the staff that services the animals.
- The number of animals required to produce the same amount of food has been vastly decreased by shortening the animals' lives.
- CAFOs are generally owned by large corporations that can procure supplies and other items from anywhere in the world and can use their purchasing power to negotiate the best terms.
- In the United States and most Western countries, the production of grains is subsidized by the government, thus reducing the price of feed.
- CAFOs are treated as both agricultural and industrial establishments in the United States and hence enjoy the tax benefits of both types of establishments.

The industry is able to get away with practices that adversely affect the environment because such damage is not immediately perceptible. While the economic benefits of activities like cutting down forests, intensive growing of the same crop over and over again, overuse of groundwater, and dousing farm animals in chemicals are immediate, their environmental effects become apparent only over time. Rules and regulations set up by governmental authorities, such as the EPA in the United States, are generally designed to prevent egregious acts of pollution like spills from manure pits into waterways and do not address the question of long-term sustainability of the present mode of operation. Even the regulations that are in place are often violated, because the risk of substantial financial penalties is minimal. There is invariably a literal thick wall around industrial livestock operations—erected in the name of avoiding contamination—that also facilitates secrecy and insulates the public from their facilities.

Animal factories are generally set up in regions with depressed economies and lax environmental regulations; the attraction of jobs and other economic benefits encourages local governments to invite the factories to their region. The region has to provide land, water, access, power, and other inputs that are needed for the operation of CAFOs. Meanwhile, the promoters of CAFOs promise to enhance the economic base of the community through expenditures in building, equipment, feed, and jobs—promises that are stated or implied but never written, and that, as research has consistently shown, do not materialize. Since CAFOs are owned by large corporations with access to national and international markets, they acquire necessary items and supplies from outside the region, not from within it, and the employment they provide consists of low-paying jobs. Even in cases where large farming operations bring more jobs and total income to communities, they also bring greater inequity to income distribution. Most of the promised income never materializes, and CAFOs spend relatively little for operating needs within the local communities.

The two most important resources that are consumed in large quantities, either directly or indirectly, are water and fossil fuels. There is already a shortage of water—a resource that is crucial for our survival—in many parts of the world. As we have seen, producing foods of animal origin requires much more water than producing agricultural products for consumption by humans. Arjen Hoekstra of the University of Twente and Ashok Chapagain of the World Wide Fund for Nature have calculated that a slice of bread, a slice of bread with cheese, and a glass of milk are made at the expenditure of 40, 90, and 200 liters of water, respectively. It takes 135 liters of water to produce an egg, 170 liters of water to produce a glass of orange juice, and 190 liters of water to produce a glass of apple juice. However, all of this water pales in comparison with the 2,400 liters of water required to produce a hamburger.[9] As mentioned in chapter 6, an omnivorous diet requires 5 to 10 times more water than one consisting of plant products. The water used in producing animal food items is either returned in a polluted form or evaporates into the atmosphere. Disregarding all other environmental degradation, and including only the true cost of water, meat is prohibitively expensive.

ALTERNATIVE APPROACHES: ORGANIC AND LOCAL FARMING

Substantially reducing the consumption of animal-based foods will begin to reduce the burden on the environment in a short time, with a cumulative effect that will grow with the passage of time. Decreasing the adverse effects of the industrial production of meat on the environment can be done in other ways, too, such as buying organic food and produce from local small-scale farms. An organic farming system relies on ecologically based practices including biological pest management, not using synthetic chemicals in crop production, and prohibiting the use of antibiotics, insecticides, pesticides, and hormones in livestock production. Organic farming preserves the fertility of the soil by preventing the rapid loss of topsoil and consuming smaller quantities of fossil fuels. Organic livestock farms also attempt to accommodate the animals' natural, nutritional, and behavioral requirements by giving them suitable living conditions and allowing them access to the outdoors.

Fields that use the organic method currently have yields that are lower than conventional farms by 30 or 40 percent.[10] These results do not mean that it is not possible to obtain yields that are comparable to those of conventional farms.[11] A study carried out over 21 years found that, although crop yield was 20 percent lower on organic farms, the input of fertilizer and energy was reduced by 34 to 53 percent. Enhanced biodiversity and soil fertility on organic farms were due to the use of compost and legume-based rotation of crops.[12]

According to data maintained by the USDA, most of the organic farms in the country are small, typically less than 10 acres.

The basic approach of organic farming is substantially different from that of conventional farms. Labor requirements on organic farms are slightly greater than on conventional farms, and there is an important difference in the labor, in that organic farming requires a knowledgeable staff that makes decisions based on the local conditions as they change over time. On large conventional farms, the prescriptions for the application of fertilizers and chemicals are provided by the company that sells the seeds, and the entire process is carried out with the help of mechanical equipment. Workers on organic farms, on the other hand, must be intimately familiar with changes in local conditions so they can take corrective actions as necessary. While the acreage of organic farms in the United States is about one-half of 1 percent of the total, the proportion of organic cattle is even smaller. According to the USDA, there were only 200,000 organic cattle in the country in 2005, of a total of close to 100 million. Land requirements for the production of organic meats are roughly twice as large as for conventional farms.[13]

The demand for organic products has been increasing rapidly in all parts of the world, primarily because of concerns about the chemicals and hormones used on conventional farms. At present, there simply are not enough organic cows in the United States to meet the demand, even disregarding the fact that there is not enough organic grain to feed them. During the last decade, the sale of organic products has increased at the rate of more than 10 percent each year, with no sign of leveling off. The price of organic products in the market is usually higher than that of conventional products by a substantial amount, even a factor of two or three. It may be that the price differential will decrease if organic farming becomes very popular.

The demand for organic milk is increasing faster than the demand for organic meats, which is one reason that more dairy farms are dedicated to the production of milk than to meat. While organic dairy farms offer a number of advantages over CAFO style operations, primarily because they do not use chemicals that are routinely used in industrial operations, their effect on the environment is not entirely benign. The emission of various gases and particulate matter will be roughly of the same order as in CAFO operations. If hay and forage are substituted for grains to a substantial extent, the release of methane from the animals' digestive system will be even greater. Organic farms also require more land to produce the same amount of milk. When all of these factors are taken into consideration, the adverse environmental impact of organic milk is somewhat less than that of the production of conventional milk, but it is still substantial.[14]

Large corporations getting involved in organic farming brings another set of problems and concerns. It is highly likely that these companies will meet the minimum requirements to get the organic label while at the same time trying to continue doing things the old way. For example, the USDA requires that certified organic animals must have access to pasture; it does not specify how much land should be available to each animal. Thus, a conventional feedlot can obtain an organic label by providing a small enclosure next to the housing that provides limited access to pasture for the cattle for a very short period of time, and which only some of the animals may use.

Another problem emerges because developed countries regularly import fruits, vegetables, and other food items from developing nations. When there are not enough organic products in the market, organic companies import products from China, Brazil, Sierra Leone, New Zealand, and other countries.[15] Even with frequent inspections, it is not possible to ensure that all products received from foreign countries satisfy the requirements of being truly organic. For these reasons, many people believe that organic farming must remain small-scale and local to be true to its name. Whether such farms can meet the growing demand for organic products remains to be seen.

Recently there has been significant interest in meats from local farms that raise livestock. These operations do not produce the problems that are endemic in industrial operations. The manure of the animals is plowed into the soil to provide nutrients after composting, the use of chemicals is very limited, and pathogens that grow and afflict the animals in CAFOs do not have an opportunity to grow in more natural settings. On the other hand, by their very definition, these will be small-scale operations that can only cater to the needs of a limited number of people. It will be difficult for most people living in urban and semiurban areas to find an outfit that provides meat from animals raised in that region.

There is also some interest in beef from grass-fed cattle, because beef from these animals has smaller amounts of fat and includes some omega-3 fats that are found in greater proportions in grass and hay. Such beef will certainly decrease the harmful effects of a diet rich in red meats, but there is no significant advantage from an environmental point of view. Grass-fed cattle take longer to mature, and hence the resources used are not significantly less than those used in other settings. The most significant problem with this type of meat is that most such beef comes from Brazil, where rainforests are razed to feed the cattle. Over a period of seven years, the export of Brazilian beef increased by more than a factor of four, and millions of acres of rainforests were razed.[16]

VEGAN OR LACTO-OVO VEGETARIAN?

Vegetarianism is defined in many different ways. While vegans abstain from all foods of animal origin, lacto-ovo vegetarians only abstain from eating meat and fish. A vegan diet eliminates all types of environmental problems associated with raising livestock, not just those produced by animal factories. Directly consuming agricultural products reduces the waste associated with the conversion of grains to animal products, thus allowing a greater number of persons to live on the resources of a given piece of land. For the same reason, people following a vegan diet represent a smaller burden on planetary resources of water and fossil fuels. A reduction of saturated fats and cholesterol in a vegan diet helps reduce their chances of suffering from chronic diseases as compared with those consuming a typical meat-based diet. Lacto-ovo vegetarians derive the same benefits from their meals as vegans, and vegans do not appear to have any significant advantage over this group in the incidence of chronic disease.[17]

On the negative side, eliminating all types of animal products increases the risk of certain nutritional deficiencies. Micronutrients of special concern in the vegan diet are vitamins B12 and D, calcium, and long-chain omega fatty acids. Vitamin B12 is needed for cell division and blood formation. Plant foods do not contain vitamin B12 except when they are contaminated with microorganisms. For some vegans, iron and zinc status is also of concern because of the limited availability of these nutrients. Unless vegans consume foods that are fortified with these nutrients, they may need to take dietary supplements. Some vegans believe that these deficiencies are actually created by the "unnatural" food production processes in play today, processes that inhibit the availability of vitamin B12 in plant-based foods. Fresh garden vegetables and fermented foods produced in porcelain vats, like some types of tempeh and miso, are supposed to be rich sources of vitamin B12, but the industrial production of vegetables and the stainless steel vats used in the production of most tempeh and miso make these food products less reliable sources of vitamin B12.

While modern dairy farms create environmental problems similar to those produced by beef factory farms, at least some of the problems of dairy CAFOs are associated with the intensity of production and the amount of milk that a dairy cow is made to produce while being fed grains and other energy-rich items. On the basis of environmental and health-related considerations, the best approach would be to reduce the consumption of milk and dairy products by substantial amounts in developed countries. Consuming less milk, yogurt, and cheese will reduce pressure on the dairy industry to produce excessive amounts of milk.

As ruminants, cows convert fibrous plant products that are of no use to humans into a nutritious and high-value product. For example, seeds from which oil has been extracted make particularly valuable feed items because of their protein content. Agricultural waste products can provide the bulk of the feed for dairy cows. In addition to reducing the environmental burden of producing high-energy feed, giving cows this diet would also convert waste products to a high-value, nutritious food. Replacing dairy CAFOs with smaller-scale, decentralized operations would be beneficial for the environment because of reduced consumption of high-energy feed and because the manure could be plowed back into the land. Thus a large reduction in the demand for milk will be beneficial both for the environment and human health.

The main advantage of eggs is that they provide a very inexpensive source of protein and some micronutrients. This feature is particularly useful for people whose diets do not have sufficient amounts of protein, particularly in developing countries. As for energy considerations, the production of milk by dairy cows and eggs by laying hens is more efficient than the production of other types of animal products. The production of eggs also does not require copious amounts of water, thus saving this precious resource. If hens are raised in a truly free-range manner, the incidence of *Salmonella* and other pathogens would also be substantially less than in operations that use large housing chambers over and over again.

FOOD CHOICES AND WORLD HUNGER

Who can remain indifferent to the extent of hunger and malnutrition in the world? In the developing world, an estimated 174 million children under the age of 5 were malnourished in 1996-1998, and 6.6 million out of 12.2 million deaths in that age group were associated with malnutrition.[18] According to the Food and Agriculture Organization of the United Nations (FAO), the number of malnourished people has been rising continuously; the number of hungry people was estimated to be 1.02 billion in 2009. A shortage of food inevitably results in higher food prices, and food prices have been rising for the last few years. In many countries in the developing world, people spend 60 to 70 percent of their family income on food, so an increase in the price of food inevitably leads to hunger or malnutrition.

The surge in food prices in the last few years has resulted in a 50 to 200 percent increase in selected commodity prices. Food, the fundamental determinant of health and requirement for life, is unaffordable to an increasing proportion of the world's population. In the United States, the proportion of income spent on food is, for most people, less than 15 percent. Since a large part of the cost of food goes into packaging, advertising, and marketing, the overall effect on consumers of an increase in the price of grains is rather small. However, even a change this small will increase the number of persons in the developed world who cannot buy food all through the year.

The burden that a population puts on farmlands depends on the choice of food. In low-income countries, such as India, where grains supply 60 percent of calories, an average person directly consumes more than a pound of grain per day. In affluent countries, such as the United States, the consumption of grains per person is more than four times this amount, although 90 percent of it is consumed indirectly as meat, milk, and eggs from grain-fed livestock.[19] A farmer can feed up to 30 people throughout the year on 1 hectare of land with vegetables, fruits, cereals, and vegetable oils. If the same area is used for the production of eggs, milk, or meat, the number of persons fed will be between 5 and 10.[20]

Of all the types of environmental degradation and exhaustion of resources, the shortage of freshwater is likely to have an impact on the availability of food in the shortest time. The level of groundwater is falling in almost all parts of the world, including the most populous countries, China and India. Economists predict that by 2025 water scarcity will reduce global food production by 385 million tons per year, which is more than the current U.S. grain harvest.[21] The production of wheat in China, the largest producer of this grain, has decreased by 8 percent since it peaked at 123 million tons in 1997. During the same time period, rice production dropped by 4 percent.[22] Water in millions of wells dug in all the states of India has disappeared or fallen too low to be accessible with conventional methods. The countries with absolute water scarcity will have to import a substantial portion of their cereals, while those unable to finance these imports will be threatened by famine and malnutrition.

A clear indication of the shortage of food, as envisaged by many countries, is the fact that dozens of private investors and governments, including Saudi Arabia, Qatar, South Korea, and China, have leased hundreds of thousands of acres of farmland in Africa and South America to protect their home countries against the inevitable shortages. Such deals can easily be made with countries in Africa because the regional governments are not very powerful and are often attracted by money, but some negotiations are also being conducted with East Asian countries such as Thailand and Vietnam. China has invested $800 million in rice production in Mozambique, and Jordan has secured tens of thousands of hectares for raising livestock and growing crops in Sudan. Corporate investors in London and New York have shown interest in farmlands as investment vehicles.[23] Some countries are also seeking long-term trade contracts that would ensure future supplies of food in return for cash. The Philippines has signed a three-year contract with Vietnam for a guaranteed 1.5 million tons of rice each year.

In addition to various kinds of environmental degradation and exhaustion of precious resources, global warming, signs of which are abundantly clear, presents another set of problems for the availability of food. The rule of thumb among crop scientists is that for every 1°C (1.8°F) increase in temperature, the output of farms that grow rice, corn, or wheat decreases by 10 percent. The explanation is that crops have evolved over centuries to make optimum use of weather conditions. Perhaps a larger danger to the production of food is droughts or excessive amounts of rainfall—both of which have been occurring in many regions of the world, including some of the most productive areas, that provide food to billions of people. An analysis performed by the National Resources Defense Council examined the effect of global warming on water supply and demand in the contiguous United States. The study found that more than 1,100 counties—one-third of all counties in the lower 48 states—will face higher risks of water shortages by midcentury as a result of global warming.

SCIENCE TO THE RESCUE?

There are irrefutable signs of damage to the environment around us which have been documented by respected national and international organizations such as the U.N. Environment Programme, the Intergovernmental Panel on Climate Change, the Worldwatch Institute, the U.N. Millennium Project, the EPA, and the U.S. National Aeronautics and Space Administration. The pace of melting of ice caps, changes in weather patterns, depletion of fisheries, and other changes is generally greater than that predicted by these organizations. Scientists are generally conservative in making estimates, in part so they are not blamed for creating unnecessary panic.[24] The contribution of livestock operations to these changes strengthens the environmental case against the consumption of foods of animal origin.

Projections of the supply of and demand for food clearly indicate that providing the desired kinds of food to the human population during the next few decades will be a major challenge, if not impossible. However, projections of food shortages caused by environmental degradation fail to impress people who have plenty of food, albeit at a somewhat higher price than last year. One reason is that we have difficulty comprehending the gravity of a situation that lies beyond our range of perception. The second reason is that technological innovations—advances in telecommunications, computers, the Internet, space exploration, and gizmos that keep coming out at a feverish pace—have dazzled people and convinced them that science will find a solution to all of humanity's problems. Finally, previous projections of great hardships due to shortages of food were widely off the mark. The British economist Thomas Malthus predicted in the early nineteenth century that the supply of food could not keep pace with the increasing population. In 1968 Paul Ehrlich published the book *The Population Bomb,* in which he predicted dire consequences of the increasing population within a few decades. The Club of Rome, a global think tank, produced a book entitled

World population and projections, in billions.

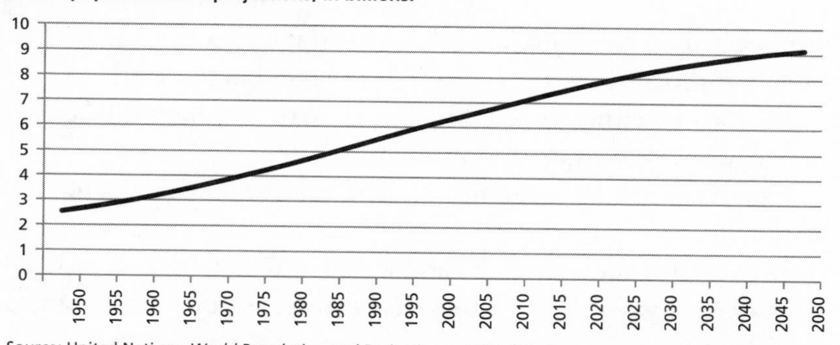

Source: United Nations, *World Population and Projection to 2050,* http://esa.un.org/unpp/index.asp?panel=2.

The Limits to Growth in 1972 with somewhat similar conclusions. The club later modified its position and supported the publication of *Mankind at the Turning Point*, which suggested that the situation in the future will be ominous but within human control.

At the time Malthus made the prediction of widespread hunger and starvation, much of the world was undiscovered and unexploited for the production of food. In the late 1960s, when Paul Ehrlich and the Club of Rome made their predictions, many countries in the world, especially India, were facing acute shortages of food, with the distinct possibility of mass starvation. However, agronomic research pioneered by Norman Borlaug and the International Rice Research Institute produced highly productive dwarf varieties of wheat, rice, and other crops that led to the Green Revolution. As a result of these developments, agricultural production not only kept pace with the demand for cereals but exceeded it. India, for example, became a major exporter of rice to Africa and the Middle East.

In addition to the agronomic research that produced hybrid varieties of grains, a number of other factors contributed to the rapid increase in global food production during the last four or five decades of the twentieth century. According to FAO statistics, global fertilizer use more than quadrupled during this period, an additional 10 percent of land on a worldwide basis was brought under cultivation, and the area of farmland that is irrigated—and hence has a much greater productivity—almost doubled. A simple extrapolation from increasing demand requires that the production of grains double again by the middle of this century; this, in turn, will require at least a threefold increase in the amount of fertilizer applied to the crops, a doubling of the irrigated areas, and an 18-percent increase in the amount of cropland in the world. These expansions are either impossible to achieve or fraught with dangers that may have consequences on a global scale.

Producing the same amount of food year after year will require major innovations in conservation and horticultural research. Doubling the production of food by 2050 will require additional areas of cultivated land—most likely obtained by cutting down rainforests—and sources of freshwater that simply do not exist. Application of increasingly greater amounts of chemicals and fertilizers will poison the land. These factors will make it increasingly difficult to produce the same amount of food year after year on a per capita basis during the coming decades, even at the present unsatisfactory level, which leaves more than a billion hungry people. Since markets are connected globally, a shortage of food will cause an inflationary spiral (which may have already begun), causing people with lesser means to suffer.

Agronomic research to continue increasing yields has reached a point of diminishing returns; we may be close to the limit of what a given biomass of plants can produce. During the last few decades, the amount of grain produced by the best varieties has remained the same; small gains in yield per acre in the production of grains have been achieved by producing strains that are less susceptible to losses of various kinds. Strains of wheat and rice produced during the Green Revolution increased the harvest index—the ratio of grain to the mass of the entire plant—to about 50 percent. It is unlikely that this number can be increased much further, because leaves and stems are required by plants to survive and to produce the grain, and there may be a physical limit to the amount of grain that a plant can produce. Breeding shorter wheat plants produces such a low, uneven canopy that it does not intercept light uniformly, thus reducing photosynthesis; breeding thinner and lighter stalks makes plants more likely to collapse under the weight of their own grain.[25] These factors do not undermine the importance of agronomic research, which must continue to breed plants tolerant of pests and adverse environmental conditions. Unfortunately, pests keep evolving to become resistant to each new innovation. For example, corn hybrids in the United States become useless after about four years, at which time they must be replaced by new strains.

Our need to increase agricultural productivity to keep pace with the projected demand will have dangerous consequences. Intensive farming and excessive application of fertilizers is decreasing the productivity of the land by causing a loss of topsoil and increasing salinity to harmful levels. Increasing the application of fertilizers will further increase damage to the soil, to intolerable levels. The only way to significantly increase the area of agricultural land is by cutting down forests. Undeveloped land and forests provide essential services like carbon sinks and habitat for numerous species of plants, insects, and animals, and they have a moderating effect on the climate. Finally, increasing the amount of irrigated land to the desired level appears to be impossible. Water is already the limiting factor for agricultural productivity; reduced sources of freshwater indicate that there may not be enough water to maintain even the present level of productivity for long.

Every indication is that there is going to be a conflict. The burgeoning human population and its changing lifestyle will outrun our planet's resources. Global warming and climate change will almost certainly adversely affect the yields of farmlands on a large scale. As a result, the ecosystem may force us to make difficult choices within the next few decades, if not sooner. It is true that scientific developments may address some of these problems and move the time frame forward by a few years. For example, plants may be genetically modified to withstand a limited amount of drought, and water conservation techniques may decrease the need for irrigation. Science's capacity to tackle factors that affect food production on a planetary level is very limited, however. If the oceans are being depleted of marine life due to overfishing in major bodies of water, there is little that scientific innovation can do to change the outcome. Even genetic modification cannot make deserts bloom or produce food crops from lands that suffer from excessive salinity, depletion of nutrients, or loss of topsoil. The scale of these problems is completely different from the scale of problems that can be addressed by new technologies and products.

SUSTAINABILITY AND EQUITY

A 1987 United Nations report defined sustainability as meeting the needs of the present generation without compromising the ability of future generations to meet their needs.[26] A pasture with a limited number of farm animals meets the criterion of sustainability if the land is able to regenerate the vegetation eaten by the livestock. Many civilizations in the preindustrial era were sustainable. The dependence of present farming methods and lifestyles on fossil fuels—an exhaustible resource—means that they are not sustainable. Although long-term projections carry an inherent uncertainty, it is reasonable to consider the state of the planet we are leaving to the next one or two generations. For this purpose, extrapolating the needs and available resources up to the year 2050 is appropriate.

Sustainability also means that the natural system should be able to recycle the waste and harmful substances generated by humans; the amount of the earth's surface needed to do so is known as the ecological imprint of humanity. The ecological footprint was 25 percent greater than the capacity of the earth in 2003 and is continuously increasing, projected to reach twice the earth's capacity by the middle of this century. By then, two earthlike planets will be needed to sustain our level of consumption and pollution.[27]

An indicator of the land's capacity to provide food is the so-called Net Primary Productivity (NPP), which measures the total vegetable matter produced in the world each year. According to U.N.'s *Global Environmental Outlook,* the NPP decreased by 12 percent from 1981 to 2003, primarily due to the loss and degradation of farmlands and the clearing of forests.[28] In the final analysis, humans depend on the NPP of the planet when they consume plant products directly or consume meat produced through plant products, with a large loss of energy in the conversion. That loss reduces the food available to humans. The part of the NPP not consumed by humans performs ecosystem services essential for our welfare and survival, such as absorbing carbon dioxide to produce oxygen, moderating climate, and providing biological material for various activities (for example, producing timber, providing sustenance to wildlife, sustaining grasses and trees in urban and semiurban settings, and absorbing pollutants). At the same time that the NPP is decreasing, the demands of the human population are putting a strain on the planetary food budget. These demands and strain will both increase over time, with the distinct possibility of a shortfall that cannot be covered by any means.

Depending on some unknown entity to prevent catastrophic events is not an option. We cannot maintain the status quo until things *really* come to a head. With shortages of food that are responsible for the undernourishment of a billion people, we have to face questions with ethical overtones—one of them dealing with people who are living on the edge today, and the other concerning future generations everywhere. Should we keep consuming foods of animal origin in large quantities that will either lead to deprivation elsewhere or indirectly affect the availability of food by inflating its price? Do we have a right to despoil the world and use resources to the extent that the options available to our children are severely limited? An extrapolation of the present trend for the availability of food in the coming decades forces us to consider these questions with a certain urgency. The choices we make today determine whether our children will have to lower their expectations and will influence whether many, many people in poorer countries will be pushed over the edge into the abyss.

Equity and social justice are considerations when we think about what we eat. In simple terms, justice means that we do not convert grains into fuel or meat while some people do not even have enough food on the table. In an international market, there are other implications of our choices. Richer people often procure the food they desire at the expense of the basic necessities of poorer people. For example, about 80 percent of the shrimp sold in the United States is raised in aquaculture farms, in Vietnam, Thailand, and the Philippines, that are mostly owned by multinational corporations, with the sole objective of providing shrimp to developed countries. These facilities take over precious land and bodies of water that were formerly used by rice farmers and fishermen to feed the local people; hence, shrimp farms deprive indigenous people of an important source of income and food. In many countries in South Asia and the Middle East, the meat of sheep and goats is highly prized, at least by those who can afford it. These animals are highly destructive to the environment, however, because they eat all kinds of plants, denuding the land. Hence the prized meat of richer people reduces the ability of poorer people to survive, whether locally or globally.

HEALTH CONSIDERATIONS

Extensive research on diet and nutrition during the last 10 to 20 years has firmly established the benefits of a vegetarian diet. A number of epidemiological studies involving thousands of persons and their eating habits and incidence of chronic disease were carried out for many years, even decades. As discussed in chapter 7, these studies have shown that fruits, vegetables, grains, nuts, and seeds provide numerous benefits and ward off chronic diseases to such an extent that they should be considered essential foods.

The common perception that meat is the best source of protein is only partly true. Most pulses and legumes have roughly the same protein content as meats, although their biological availability is somewhat less than that of meat, unless they are combined with grains. Soybeans are an exception because their protein content is much greater than that of meats, grains, or pulses. Consider, too, that the consumption of large amounts of protein may be unnecessary or even harmful.[29] The per capita consumption of protein by Americans is 112 grams, almost twice the amount recommended by the American Dietetic Association.

There are a huge number of phytochemicals in plants. There are so many that it is difficult to assign a role to each compound, and trying to designate roles may even be counterproductive, because various chemicals in these items work together to produce beneficial effects. In contrast to the positive effects of phytochemicals, meat consumption increases the risk of getting cancers of the colon, rectum, pancreas, and prostate, and the saturated fats and cholesterol in animal-based foods adversely affect the metabolic syndromes that are related to cardiovascular disease. According to the American Institute for Cancer Research, consuming 400 or more grams per day of a variety of fruits and vegetables could, without any other change in lifestyle, decrease overall cancer incidence by at least 20 percent.[30] While statistical studies have validity only for a large segment of the population, prudence requires that we minimize our personal risk. (A careful driver may be involved in an accident, and a careless driver may avoid accidents for many years, even though statistical analyses show that careless drivers will be more prone to accidents. This does not mean that we should all become careless drivers.)

Although plant foods such as whole grains, legumes, beans, nuts, fruits, and vegetables may also be consumed by omnivores, the amount they consume in side dishes may not be sufficient to provide the full range of benefits of a vegetarian diet. Since animal products are rich in nutrients such as proteins and fats, for an omnivore, grains, vegetables, and fruits must be limited to accompaniments to a meal. It is also noteworthy that eating large quantities of fruits and vegetables has been shown to help greatly in the prevention of chronic disease. In the Nurses' Health Study, the two groups that ate the most fruits and vegetables had 6.2 and 4.5 servings per day, respectively, of these two classes of foods. Intake of fruits and vegetables on the Hallelujah Diet were even greater: 22.7 ounces each day of fruits and 34 ounces each day of vegetables. Incorporating quantities like these in a diet that has meat as the main course would be very difficult.

The case for eating a vegetarian diet for its health benefits has to be based on scientifically established *advantages* of a diet rich in plant products and *not* on the harmful effects of eating meat, which are significant but smaller in comparison. A vegetarian diet may also be unhealthy, particularly if it is rich in saturated fats and lacks *variety*—the most important feature of healthy vegetarianism. Vegetarian foods such as french fries, soda, and chips—the usual accompaniments to hamburgers—may contribute more to obesity than the hamburger itself.

Statisticians and policy planners are mainly concerned with ensuring that the agricultural system is able to provide sufficient food, as measured by calories, to feed the projected global population of 9 billion in 2050. However, a healthy diet must be sufficient not just in calories but also in the balance of micronutrients, vitamins, and minerals provided by phytochemicals in order to ward off morbidity and mortality and help maintain or improve quality of life. It has been estimated that 2.6 million deaths per year can be attributed to inadequate consumption of fruits and vegetables.[31] The association between diet and health that has been illuminated as a result of extensive research during the last few decades should be put to use to prioritize health as a major consideration in the choice and production of food.

* * *

In an ideal world, people would be allowed to choose the diet they prefer. But the current situation and future trends indicate that increasing population, overconsumption, degradation of the environment, and exhaustion of precious resources may impose serious constraints on absolute dietary freedom. We must drastically reduce our consumption of foods of animal origin and eat more plant-based foods for our lifestyles to become sustainable.

EPILOGUE

Any one of the three primary arguments for drastically reducing our consumption of foods of animal origin is convincing all by itself. First, we want to make sure that our children do not suffer deprivations due to degradation of the environment and exploitation of the earth's resources far beyond sustainable levels. Second, we want to free up food for direct consumption by humans, since the conversion of plant products to animal-based foods is an intrinsically inefficient process. Third, we want to derive health benefits from the numerous phytochemicals in fruits, vegetables, nuts, and whole grains. It is easy to see that the consumption of meats of various kinds must be drastically reduced, if not eliminated altogether.

The production of milk and eggs does not adversely affect the environment as drastically as the production of meats, and the environmental effects of milk and egg production can be further reduced by adjusting our approaches to producing these items. Consuming these items in moderation—at levels much lower than the current levels of consumption in the Western world—will make the best use of the earth's limited resources and ensure that people do not suffer from nutritional deficiencies that can be associated with eliminating animal products completely.

Numerous aspects of our lives endanger the ecosystem, threatening its ability to continue to provide services needed for our sustenance and welfare. Many "things" are considered to be essential for modern lives; but we also have a fascination with the latest gizmos that our technologists keep designing and factories keep churning out, eminently aided by the mass media that equates nonpossession with deprivation. One question is, why consider food when there are so many aspects of modern life that impose a serious burden on the ecosystem? The average Westerner is consuming planetary resources at a rate that is far beyond the capacity of the earth to sustain for any length of time, and the emerging class of affluent people in developing countries is rapidly following the same pattern. However, food, particularly animal-based food, creates a greater strain on planetary resources that any other anthropogenic activity. The human population has become so large that a preference for foods of animal origin is having consequences on a global scale.

Environmental degradation above sustainable levels is caused by many aspects of modern life. There is an increasing awareness that some actions are necessary to minimize their impact, even as we search for lasting solutions. However, the average person cannot immediately switch to mass transit, start driving an energy-efficient car, or move to an energy-efficient home. Even someone who accepts the premise behind these changes may not be able to do these things or at least must plan to do them over years or decades. The advantage to the environment of these approaches may not be perceptible for a very long time—too long, in view of the rate at which changes are taking place.

The magnitude of the beneficial effect of switching to a vegetarian diet is at least as great as any of these items, and a change in diet will have an effect within weeks, months, or one to two years. A broiler chicken is slaughtered at the age of six weeks, pigs reach market weight in a few months, and cattle are ready for slaughter less than two years after birth. Hence switching to almost all plant-based foods will have an effect in a very short time, because these animals will not be needed and therefore will not be raised. It is also useful to remember that farm animals are completely dependent on us—we determine the number born and how long they will live. A change in eating pattern will soon have an impact on the number of livestock raised.

Arguments in favor of vegetarianism based on health considerations have a universal appeal. Chronic diseases not only decrease our lifespan but also adversely affect our quality of life for a substantial period and even severely limit our activities. Data collected in developed countries have shown that vegetarians have a lower BMI, indicating that obesity is less of a problem among vegetarians. In general, vegetarian diets have a much greater variety than meat-based diets, and thus they are easy to follow for long periods by switching among the multitude of dishes available.

The arguments in favor of vegetarianism are so strong that one wonders why the number of vegetarians in the United States and developed countries is only 3 percent. One would think that such reasoning should appeal to a vast number of people. The desire to remain healthy and not be afflicted by chronic diseases has a great appeal. All of us genuinely want to see our children at least as well off as we are, if not better, and hence sustainability of the environment should be important to everyone. A genuine concern for people who may be deprived of basic necessities, even those who live in distant places, is also built into us. We provide billions of dollars in charity to secure food, clothing, and medicine to people everywhere, and a natural disaster even in remote places elicits a sympathetic reaction.

The simplest reason for the small number of vegetarians is a lack of knowledge and information. Even basic details of the livestock industry are unknown to most of us—the number of animals that have to be raised to meet our demand for animal products, the manner in which animals spend their lives in CAFOs, the resources needed to produce these items, and the overall impact on the environment. Food, particularly animal-based food, is just too familiar and plentiful to deserve serious consideration. The amount of water needed to produce a pound of beef will surprise almost everyone, and knowledge of the precarious situation of water in all parts of the world gets buried under the details of numerous daily events. Water is something that we have not learned to take seriously, perhaps because the faucets in our homes appear to be connected to an unlimited source of water. Making a compelling connection between the consumption of a basic food item that is available almost everywhere and environmental degradation on a global scale is not easy.

The commercial undertakings that produce and bring meat to the table represent perhaps the largest industrial establishment in the country. They include farmers who grow immense amounts of corn; various chemical companies that produce fertilizers, antibiotics, drugs, and other chemicals; ranchers who work by contract to graze cattle; multinational corporations that control the operations of CAFOs; transportation companies that move animals, their feed, and other materials within and across national boundaries; producers and distributors of meat; and supermarkets. There is bound to be opposition from these vested interests against a significant shift toward vegetarianism. The vast economic clout of this conglomerate can allow it to project a favorable image in the media and also to muzzle any significant opposition.

While considering opposition to vegetarianism by entrenched interests, it is useful to think of groups that may be more receptive to these ideas. Generally speaking, rich people in the developing world tend to consume meat in greater quantities than those who do not have enough resources, partly because wealthier people often develop a taste for things that are not readily available to everyone. As a result, in developing countries, this section of society suffers from ailments that are common in developed countries, such as obesity and chronic diseases. Things are almost the opposite in the developed world. Perhaps the cheapest foods that one can find in the United States are hot dogs, hamburgers, and soda. These items can easily provide more than the sufficient amount of protein, fat, and calories. Statistical surveys have shown that poorer people do not buy many fruits and vegetables,[1] and there is a greater incidence of obesity and chronic diseases among them. On the other hand, wealthier and more educated persons tend to be more health conscious and buy leaner cuts of meat and more fruits and vegetables. It is easier to convince people who are already making health-based decisions about food of the virtues of making the quantum leap to vegetarianism, particularly when one adds sustaining the environment and alleviating world hunger to the mix.

A few studies have shown that 68 percent of vegetarians are women and 15 percent of college students in dining halls request vegetarian meals.[2] Younger persons, particularly women, are more receptive to vegetarianism because all of its associated values—preservation of the environment (both for its own sake and for future generations), better health, even compassion for animals—have a greater appeal to them. These data show that the target audience for propagating vegetarianism should initially consist of educated people. Once this group accepts the benefits of, and the logic behind, vegetarianism, the ideas will percolate to other sections of society. Although it is useful to have a target audience that is more receptive to the ideas, it would be unwise to ignore any segment of society. People everywhere are concerned about the future of their children; the important thing is to convince them that the dangers will affect not their great-grandchildren, or some generation in the distant future, but the children who are playing in their backyard right this minute.

Some people bring into discussions of vegetarianism the natural tendency of primates to consume meat or even speculations based on the shapes of our teeth. Such arguments are distractions that do not address the problems that humanity as a whole is facing. Besides, monkeys and apes are primarily vegetarians; meat constitutes only about 3 percent of their diet. Arguments based on the design of our teeth are specious. Our teeth have a much greater resemblance to those of herbivores than carnivores. In addition, as intelligent, civilized beings, we have the power to examine the ramifications of our actions, particularly in the interest of future generations, instead of appealing to the animal instinct in all of us. It is difficult to find many instances in which civilized people are champions of primitive instincts.

The industries that provide meat to us, from the farmers to the slaughterhouses, will have to make adjustments. To some extent, this burden can be eased by the subsidies that the government is already giving to the farmers. The beneficial effects of this change will become apparent in a short time. A substantial shift in dietary pattern toward vegetarian diets would reduce the environmental problems simply by reducing livestock populations and their demand for land and other resources. On a per capita basis, the land requirements of plant-based agricultural economies are only a fraction of those with high rates of meat production. With fewer animals to feed, it might be possible to rebuild world grain reserves, ensuring dependable supplies for direct human consumption in countries facing scarcity, and to bring the price of food down in general. Reducing land use by cutting meat production would also be a very effective way to ensure that wilderness areas are maintained and even expanded. Wilderness is crucial to providing biological diversity, climate control, and a storehouse of carbon. Many collateral health gains would accrue from these changes, undertaken to stabilize the world's climate and secure our future, including a healthier diet, improved air quality, and more reliable freshwater supplies.

Some people may argue that fast foods, most of which are animal products, have freed women from excessive work in the kitchen. This argument is only partially correct. If vegetarianism becomes popular, companies will spring up to provide healthy foods that can either be eaten as prepared or prepared by others with minimal effort. Global warming, trans fats, excessive caloric intake, chemical additives, and other dangers usually strike a chord among the public and often force industry to change its ways. Industry will respond to the demand of the general population. The only danger is that strictly following the letter of the law may dilute or negate the things that are truly beneficial. For example, when studies established that oats help in lowering cholesterol, manufacturers of snack items boasted that their snacks contain oats—but many of these snacks have more calories than candy.

There is no need to advocate a doctrinaire approach that imposes a taboo against meats under all circumstances. However, we have gone past the era when minimal changes will be sufficient. While it is highly desirable that most people drastically reduce their consumption of meats, it is also useful to have a few committed people as agents of change who can become vanguards of a movement. Following a mostly vegetarian diet should not be taken as a deprivation but as something that is being done for one's own health. It is also acceptable to approach vegetarianism in a stepwise manner, with a celebration of each step, instead of as a guilt trip for not being completely successful. Environmental sustainability for the sake of our own children makes it even more palatable. A vegetarian diet as a component of a healthy lifestyle can not only increase longevity but, more importantly, enhance the quality of life for a substantial period.

Another reason to propagate these ideas is an intrinsic faith that people everywhere try to make decisions that are expected to be better for them, their children, and even the rest of the world. Ideally, people should be allowed to choose the diet that they want, but due to the pressures imposed by overpopulation, resource scarcity, and overconsumption, absolute dietary freedom is a luxury. Flexibility in the diet is necessary for reducing the demand for food, especially as humans attempt to transition to sustainable societies. Although giving up meat in favor of a vegetarian diet sounds extreme, the serious challenge we face requires extreme actions. Grim scenarios for the future, projected by respected organizations, consider the demand for food to be fixed and strategize about options for increasing the supply of various food items, an approach that assumes that food preferences and patterns are inflexible. Such long-range forecasts assume the demand for meat to be static and then try, almost always without success, to stretch the resources of the planet to meet the demand. This need not be the case; people are sensible and sensitive to the course of events and will change if they are aware of the outcomes of their actions. In fact, there is evidence that information and knowledge eventually change attitudes. As a result of extensive educational campaigns, the consumption of veal has decreased in the United States, as has the use of fur in garments. Many European countries have already banned some egregious practices of the livestock industry.

Agricultural practices during the next few decades—primarily determined by our food choices—will shape, perhaps irreversibly, the surface of the earth, including the species of plant and animal life that inhabit it and the duration for which our planet will provide the resources to sustain us. With the burgeoning human population and our opulent lifestyle, of which food is an important component, humanity's impact on the planet is pervasive. A concerted effort on the part of individuals and organizations is required to avoid the dangers that threaten our very existence. Fortunately, what will vastly improve the sustainability of the planet and future generations will do the same for our own health and well-being.

NOTES

INTRODUCTION

1 "Food Availability: Spreadsheets," U.S. Department of Agriculture, www.ers.usda.gov/Data/FoodConsumption/FoodAvailSpreadsheets.htm.

2 J. Kearney, "Food Consumption Trends and Drivers," *Philosophical Transactions of the Royal Society B* 365(2010): 2793–2807.

3 "Fossil Fuels," National Petroleum Refiners Association, www.energy.gov/energysources/fossilfuels.htm.

4 J. Woods et al., "Energy and the Food System," *Philosophical Transactions of the Royal Society B* 365(2010): 2991–3006.

5 N. McLaughlin et al., "Comparison of Energy Inputs of Inorganic Fertilizer and Manure Based Crop Production," *Canadian Agricultural Engineering* 42(2000): 21–27.

6 Nikos Alexandratos, ed., "Fertilizers and Plant Protection Agents," U.N. Food and Agriculture Organization, 1995, www.fao.org/docrep/v4200e/V4200E0m.htm.

7 *Agricultural Statistics 2001,* U.S. Department of Agriculture, Washington, DC.

8 "1.02 Billion Hungry People: One Sixth of Humanity Undernourished—More Than Ever Before," U.N. Food and Agriculture Organization, www.fao.org/news/story/0/item/20568/icode/en.

9 "Prevalence of Diabetes and Impaired Fasting Glucose in Adults—United States, 1999-2000," *Morbidity and Mortality Weekly Report*, 52(2003): 833–837.

10 J. Sabaté, "The Contribution of Vegetarian Diets to Health and Disease: A Paradigm Shift," *American Journal of Clinical Nutrition* 78(2003): 502S–507S.

CHAPTER 1. FARMS

1 World Resources Institute, www.earthtrends.wri.org.

2 Robert Louis Stevenson, *A Child's Garden of Verses and Underwoods* (New York: Current Literature Publishing, 1906).

3 Kevin Stubbs and Karen Cathey, "CAFOs Feed a Growing Problem," *Endangered Species Bulletin,* Jan/Feb1999, 24(1): 14–15, www.fws.gov/endangered/bulletin/99/01-02/14-15.pdf.

4 Danielle Nierenberg, ed. "Rethinking the Global Meat Industry," in *State of the World 2006* (New York: W. W. Norton, 2006), 26.

5 The National Creutzfeldt-Jakob Disease Surveillance Unit, www.cjd.ed.ac.uk/vcjdworld.htm.

6 J. B. Russell and Jennifer Rychlik, "Factors that Affect Rumen Microbiology," *Science* 292(2001): 1119–1122.

7 B. E. Rollin, *Farm Animal Welfare: Social, Bioethical, and Research Issues* (Ames: Iowa University Press, 1995), 290.

8 Mark Bittman, "Rethinking the Meat-Guzzler," *New York Times,* Jan 27, 2008, WK1.

9 "Chicken: Arsenic and Antibiotics," *Consumer Reports,* Jul 2007, www.consumerreports.org/cro/food/food-safety/animal-feed-and-food/animal-feed-and-the-food-supply-105/chicken-arsenic-and-antibiotics/.

10 Douglas Gansler, "A Deadly Ingredient in Chicken Dinner," *Washington Post,* Jun 26, 2009, www.washingtonpost.com/wp-dyn/content/article/2009/06/25/AR2009062503381.html.

11 "Dirty Birds: Even Premium Chickens Harbor Dangerous Bacteria," *Consumer Reports,* Jan 2007, 20–23.

12 Leland Swenson, president of the National Farmers Union, testimony before the House Judiciary Committee, Sep 12, 2000, commdocs.house.gov/committees/judiciary/.../hju67334_0f.htm.

CHAPTER 2. ENVIRONMENT

1 U.S. Department of Agriculture, http://nrcs.usda.gov/technical.land/pubs/livestockfarm.html.

2 R. Marks, *Cesspools of Shame: How Factory Farm Lagoons and Sprayfields Threaten Environmental and Public Health,* Natural Resources Defense Council and Clean Water Network, 2001.

3 D. Osterberg and D. Wallinga, "Determinants of Rural Health," *American Journal of Public Health* 94(2004): 1703–1708.

4 Richard J. Dove, Waterkeeper Alliance, Senate Committee on Government Affairs, Mar 13, 2002, hsgac.senate.gov/031302dove.htm.

5 R. Marks, *Cesspools of Shame: How Factory Farm Lagoons and Sprayfields Threaten Environmental and Public Health,* Natural Resources Defense Council and Clean Water Network, 2001.

6 H. Steinfeld et al, *Livestock's Long Shadow,* U.N. Food and Agriculture Organization, 2006, 144.

7 R. Marks, *Cesspools of Shame: How Factory Farm Lagoons and Sprayfields Threaten Environmental and Public Health,* Natural Resources Defense Council and Clean Water Network, 2001.

8 D. Nierenberg, "Happier Meals: Rethinking the Global Meat Industry," *Worldwatch Institute Paper* 171(2005), 30.

9 H. Steinfeld et al., *Livestock's Long Shadow,* U.N. Food and Agriculture Organization, 2006, 147.

10 *National Water Quality Inventory: 1998,* U.S. Environmental Protection Agency, Report to Congress, 2000.

11 Merritt Frey et al., *Spills and Kills: Manure Pollution and America's Livestock Feedlots,* Izaak Walton League of America and Natural Resources Defense Council, 2000, 1, www.nrdc.org/water/pollution/cesspools/cesspools.pdf.

12 U.S. Environmental Protection Agency, www.epa.gov/npdes/pubs/finafast.pdf.

13 *Agricultural Chemical Usage 2006,* U.S. Department of Agriculture, http://usda.mannlib.cornell.edu/current/AgriChemUsLivestock/AgriChemUsLivestock-05-23-2007.pdf.

14 H. Steinfeld et al., *Livestock's Long Shadow,* U.N. Food and Agriculture Organization, 2006, 140.

15 U.S. Centers for Disease Control and Prevention, www.cdc.gov/mmwr/preview/mmwrhtml/ss5510a1.htm.

16 Amy Chaplain et al., "Airborne Multidrug Resistant Bacteria Isolated from a Concentrated Swine Feeding Operation," *Environmental Health Perspectives* 112(2004): 137–142.

17 *Hogging it: Estimates of Antimicrobial Abuse in Livestock,* Union of Concerned Scientists, 2001, www.ucsusa.org/food_and_environment.

18 D. Nierenberg, "Happier Meals: Rethinking the Global Meat Industry," *Worldwatch Institute Paper* 171(2005): 17.

19 *Hogging it: Estimates of Antimicrobial Abuse in Livestock,* Union of Concerned Scientists, 2001, www.ucsusa.org/food_and_environment.

20 T. R. Callaway et al., "Forage Feeding to Reduce Preharvest *Escherichia coli* in Cattle," *Journal of Dairy Science* 86(2003): 852–860.

21 J. B. Russell, F. Diez-Gonzalez, and G. N. Jarvis, "Effects of Diet Shifts on *Escherichia coli* in Cattle," *Journal of Dairy Science* 83(2000): 863–873.

22 N. Planck, "Leafy Green Sewage," *New York Times,* Sep 21, 2006, 21.

23 H. Steinfeld et al., *Livestock's Long Shadow,* U.N. Food and Agriculture Organization, 2006, 142.

24 D. Nierenberg, "Happier Meals: Rethinking the Global Meat Industry," *Worldwatch Institute Paper* 171(2005): 45.

25 H. Steinfeld et al., *Livestock's Long Shadow,* U.N. Food and Agriculture Organization, 2006, 103.

26 W. L. Chemeides et al., "Growth of Continental Scale Metro-algo-plexes, Regional Ozone Pollution, and World Food Production," *Science* 264(1994): 74–77.

27 H. Steinfeld et al., *Livestock's Long Shadow,* U.N. Food and Agriculture Organization, 2006, xxi.

28 Ibid., 82.

29 Ibid., 272.

30 Ibid., 90–93.

31 Ibid., 97.

32 Gidon Eshel and Pamela Martin, "Diet, Energy and Global Warming," *Earth Interactions,* Apr 2006, 10(6): 1–17.

CHAPTER 3. LAND

1 "The Rangelands of Arid and Semi-arid Areas," International Fund for Agricultural Development, www.ifad.org/lrkm/theme/range/arid/arid_2.htm.

2 "Desertification and Arid Zones," U.N. Educational, Scientific and Cultural Organization, www.unesco.org/bpi/pdf/memopbi40_desertification_en.pdf.

3 H. Steinfeld et al., *Livestock's Long Shadow: Environmental Issues and Options,* U.N. Food and Agriculture Organization, 2006, 73.

4 G. C. Daily, "Restoring Value to the World's Degraded Lands," *Science* 269(1995): 352–354.

5 J. Robins, *The Food Revolution* (Berkeley: Conari Press, 2001), 251.

6 G. Wuerthner and M. Mattesen, *Welfare Ranching: The Subsidized Destruction of the American West* (Washington, DC: Island Press, 2002), 44.

7 "Feeding the World: Disappearing Land," World Resources Institute, www.wri.org/publication/content/8426.

8 D. Hinrichsen, "Feeding a Future World," *People and Planet,* 1998, 7(11): 6–9.

9 U.N. Committee to Combat Desertification, http://unccd.int.

10 P. Rogers, "Report: Grazing is Costly to Taxpayers," *Mercury News* (San Jose), Oct 24, 2002, www.bayarea.com/mld/mercurynews/news/local/4356709.htm.

11 "China to Restrict Grazing on Natural Grassland," Xinhua News Agency online, Sep 11, 2004.

12 L. Brown, "China Losing War with Advancing Deserts," *Earth Policy News,* Aug 5, 2003, www.earth-policy.org/Updates/2006/Update61.htm.

13 "Message on World Environment Day," U.N. Environment Programme, Jun 5, 2006, www.un.org/popin/fao/land/land.html.

14 U.S. Geological Survey, http://wfc.usgs.gov.

15 J. Belsky et al., "What the River Once Was: Livestock Destruction of Western Waters," in G. Wuerthner and M. Matteson, eds., *Welfare Ranching: The Subsidized Destruction of the American West* (Washington, DC: Island Press, 2002), 179–182.

16 Yearbook tables, Feedgrain Database, U.S. Department of Agriculture, Economic Research Service, www.ers.usda.gov/Data/feedgrains/FeedYearbook.aspx.

17 L. Starke, ed., *State of the World 2006* (New York: W. W. Norton, 2006), 30.

18 R. Lal, "Soil Erosion Impact on Agronomic Productivity and Environment Quality," *Critical Reviews in Plant Sciences* 17(1998): 319–464.

19 D. Pimentel and N. Kounang, "Ecology of Soil Erosion in Ecosystems," *Biosystems* 1(1998): 416–426.

20 Ibid.

21 D. Tilman et al., "Agricultural Sustainability and Intensive Production Processes," *Nature* 418(2002): 671–677.

22 *Global Environmental Outlook Year Book,* U.N. Environment Programme, GEO-4, 2007.

23 Nelleman et al., eds., *The Environmental Food Crisis,* U.N. Environment Programme, 2009.

24 *Global Environmental Outlook Year Book,* U.N. Environment Programme, GEO-4, 2007; Intergovernmental Panel on Climate Change, www.ipcc.ch/publications_and_data/ar4/wg2/en/ch13s13-2-4.html.

25 D. Kaimowitz et al., "Hamburger Connection Fuels Amazon Destruction: Cattle Ranching and Deforestation in Brazil's Amazon," Center for International Forestry Research, www.cifor.cgiar.org/publications/pdf_files/media/Amazon.pdf.

26 "Meat: Now It's Not Personal!," *World Watch Magazine* Jul/Aug 2004, 12–20.

27 D. Nierenberg, "Happier Meals: Rethinking the Global Meat Industry," *Worldwatch Paper* 171, Sep 2005, 32.

28 D. Pimentel and M. Pimentel, eds., *Food, Energy, and Society* (Boulder: University Press of Colorado, 1996), 173.

29 H. Steinfeld et al., *Livestock's Long Shadow: Environmental Issues and Options,* U.N. Food and Agriculture Organization, 2006, 185.

30 C. Biggelaar, "Sustainable Development," *Advances in Agronomy* 81(2004): 1–48; C. Biggelaar et al., "Sustainable Land Management," *Advances in Agronomy* 81(2004): 49–95.

CHAPTER 4. WATER

1 M. Rosengrant, X. Cai, and S. A. Cline, *World Water and Food to 202* (Washington, D.C.: International Food Policy Research Institute, 2002).

2 D. Pimentel and M. Pimentel, eds., *Food, Energy, and Society* (Boulder: University Press of Colorado, 1996), 159.

3 C. Nelleman et al., eds., *The Environmental Food Crisis,* U.N. Environment Programme, 2009.

4 Glenn Schaible, "Western Irrigated Agriculture," U.S. Department of Agriculture, Economic Research Service, Agricultural Statistics, Jul 20, 2004, www.ers.usda.gov/Data/WesternIrrigation/.

5 *Year Book 2010,* U.N. Environment Programme, www.unep.org/pdf/year_book_2010.pdf.

6 Ximing Cai and Mark W. Rosegrant, "World Water Productivity: Current Situation and Future Options," in *Water Productivity in Agriculture: Limits and Opportunities for Improvement,* ed. J. W. Kijne, R. Barker, and D. J. Molden (Cambridge, MA: CABI Publishing, 2003), 163–178.

7 D. Seckler, R. Barker, and U. Amarsinghe, "Water Scarcity in the Twenty-first Century," *International Journal of Water Resources Development* 15(1999): 29–42.

8 "Water Conservation," U.S. Environmental Protection Agency, Nov 19, 2010, www.epa.gov/oaintrnt/water/.

9 Robert Monroe and Mario Aguilera, "Lake Mead Could Be Dry by 2021," Scripps Institute of Oceanography, Feb 12, 2008, scrippsnews.ucsd.edu /Releases/?releaseID=876.

10 "Australia's Plan to Restore Parched Rivers," *New Scientist,* Jan 25, 2007, 13.

11 Felicity Barringer, "Water Use in Southwest Heads for a Day of Reckoning," *New York Times,* Sep 27, 2010, A14.

12 F. Pearce, *When the Rivers Run Dry* (Boston: Beacon Press, 2006), 185.

13 Lester Brown, "Plan B 3.0: Mobilizing to Save Civilization," Earth Policy Institute, May 2, 2009, 68.

14 Alejandro Guevara-Sanginés, "Water Subsidies and Aquifer Depletion in Mexico's Arid Regions," Human Development Report 2006, U.N. Development Program, 2006.

15 V. McGuire, *Water Level Changes in High Plains Aquifer, Predevelopment to 2007,* U.S. Geological Survey, Scientific Investigation Report, 2009-5019.

16 M. Guru and J. Horne, *The Ogallala Aquifer,* Kerr Center for Sustainable Agriculture, 2000, www.kerrcenter.com/publications/ogallala_aquifer.pdf.

17 V. McGuire, *Water Level Changes in High Plains Aquifer, Predevelopment to 2007,* U.S. Geological Survey, Scientific Investigation Report, 2009-5019.

18 Fracturing is the breaking up of rocks that support the land. It may create cracks without subsidence. D. L. Galloway, D. R. Jones, and S. E. Ingebritsen, "Land Subsidence in the United States," U.S. Geological Service, Fact Sheet 165-00, Dec 2000, http://water.usgs.gov/ogw/pubs/fs00165/.

19 Steinfeld et al., *Livestock's Long Shadow: Environmental Issues and Options,* U.N. Food and Agriculture Organization, 2006, 132.

20 "China's Rapidly Growing Meat Demand: A Domestic or an International Challenge?," Center for World Food Studies, Dec 2005, www.sow.vu.nl /Brief%20Feed%20for%20China.pdf.

21 Fred Pearce, "Virtual Water," Jun 19, 2008, www.forbes.com.

22 A.Y. Hoekstra and A. K. Chapagain, "Water Footprint of Nations: Water Use by People as a Function of Their Consumption Pattern," *Water Resources Management* 21(2007): 35–48.

23 "China's Rapidly Growing Meat Demand: A Domestic and an International Challenge," Center for World Food Studies, Dec 2005, www.sow.vu.nl /Brief%20Feed%20for%20China.pdf.

24 Fred Pearce, *When the Rivers Run Dry* (Boston: Beacon Press, 2006), 306.

CHAPTER 5. FISH

1 D. Pimentel and M. Pimentel, eds., *Food, Energy, and Society* (Boulder: University Press of Colorado, 1996), 85.

2 D. Pauly et al., "Towards Sustainability in World Fisheries," *Nature* 418(2002): 689–695.

3 D. Pauly et al., "Fishing Down Marine Food Webs," *Science* 279(1998): 860–863.
4 Maria L. La Ganga, "Federal Officials Ban Salmon Fishing off California Coast," *Los Angeles Times,* Apr 9, 2009, 1.
5 Amy Julia Harris, "Salmon Season Opens with a Whimper," *Half Moon Bay Review,* Jul 8, 2010, www.hmbreview.com/articles/2010/07/08/news/doc-4c34bb786049b498936245.txt.
6 Boris Worm et al., "Impacts of Biodiversity Loss on Ocean Ecosystem Services," *Science* 314(2006): 787–790.
7 Ibid.
8 E. Norse and L. Watling, "Dragnet—Bottom Trawling, the World's Most Severe and Extensive Seafloor Disturbance," www.mcbi.org/what/AAASsymposia.htm.
9 *Global Environmental Outlook Year Book,* U.N. Environment Programme, GEO-4, www.unep.org/geo/geo4.asp.
10 L. Reijinders and S. Soret, "Quantification of the Environmental Impact of Different Dietary Protein Choices," *American Journal of Clinical Nutrition* 78(2003): 664S–668S.
11 *Global Environmental Outlook Year Book,* U.N. Environment Programme GEO-4, www.unep.org/geo/geo4.asp.
12 R. J. Diaz and T. Rosenberg, "Spreading Dead Zones and Consequences for Marine Ecosystems," *Science* 321(2008): 926–929.
13 "Hypoxia in the Northern Gulf of Mexico," Louisiana University Marine Consortium, 2010, www.gulfhypoxia.net/Overview/.
14 *The State of World Fisheries and Aquaculture 2008,* Fisheries and Aquaculture Department, Food and Agriculture Organization of the United Nations, 2009, 21; U.N. Food and Agriculture Organization, www.fao.org/docrep/007/y5600e/y5600e.HTM.
15 V. Smil, *Feeding the World: A Challenge for the Twenty-First Century* (Cambridge: MIT Press, 2000), 178.
16 Ibid., 161.
17 K. Powell, "Fish Farming: Eat Your Vegetables," *Nature* 426(2003): 378–379.
18 R. Naylor et al., "Effect of Aquaculture on World Fish Supplies," *Nature* 405(2004): 1017–1024.
19 R. Hites et al., "Global Assessment of Organic Contaminants in Farmed Salmon," *Science* 303(2004): 226–229.
20 C. Nelleman et al., eds., *The Environmental Food Crisis,* U.N. Environment Programme, 2009.
21 Michael L. Weber, "What Price Farmed Fish: A Review of the Environmental and Social Costs of Farming Carnivorous Fish," SeaWeb Aquaculture Clearinghouse, 2003, 8.

CHAPTER 6. RESOURCES

1 V. Smil, *Feeding the World: A Challenge for the Twenty-first Century* (Cambridge, MA: MIT Press, 2000), 157–158.

2 Ibid.

3 L. Reijnders and S. Soret, "Quantification of the Environmental Impact of Different Dietary Choices," *American Journal of Clinical Nutrition* 78(2003): 664S–668S.

4 D. Pimentel and M. Pimentel, "Sustainability of Meat-based and Plant-based Diets and the Environment," *American Journal of Clinical Nutrition* 78(2003): 660S–663S.

5 V. Smil, *Feeding the World: A Challenge for the Twenty-first Century* (Cambridge, MA: MIT Press, 2000), 159.

6 L. Reijnders and S. Soret, "Quantification of the Environmental Impact of Different Dietary Choices," *American Journal of Clinical Nutrition* 78(2003): 664S–668S.

7 D. Pimentel and M. Pimentel, *Food, Energy and Society* (Boulder: University Press of Colorado, 1996); D. Pimentel et al., "The Potential for Grass-fed Livestock: Resource Constraint," *Science* 207(1980): 843–848.

8 Davis Bennett, "It Takes a Lot of Water to Grow a Corn Crop," South East Farm Press, Dec 28, 2007, http://southeastfarmpress.com/it-takes-lot-water-grow-corn-crop.

9 D. Pimentel and M. Pimentel, *Food, Energy and Society* (Boulder: University Press of Colorado, 1996), 157.

10 Ed Ayers, "Will We Still Eat Meat?," *Time,* Nov 8, 1999, 77.

11 D. Pimentel et al., *Water Resources, Agriculture and the Environment,* Cornell University College of Agriculture and Life Sciences, Report 04-1, Jul 2004, http://ecommons.library.cornell.edu/bitstream/1813/352/1/pimental_report_04-1.pdf.

12 A. Hoekstra and A. Chapagain, *Globalization of Water: Sharing the Planet's Freshwater Resources* (Malden, MA: Blackwell Publishing, 2008), 15; A. Hoekstra and A. Chapagain, "Water Footprint of Nations: Water Use by People as a Function of Their Consumption Pattern," *Water Resource Management* 21(2007): 35–48.

13 "A Fresh Approach to Water," editorial, *Nature* 452(2008): 253.

14 "The FAO Price Index," Jun 2008, www.fao.org/docrep/010/ai466e/ai466e16.htm.

15 C. Nelleman et al., eds., "The Environmental Food Crisis," U.N. Environment Programme, Feb 2009, 6.

16 *The State of Food and Agriculture 2008,* U.N. Food and Agriculture Organization, 2008, 102, ftp://ftp.fao.org/docrep/fao/011/i0100e/i0100e00.pdf.

17 Mark Lynas, "How the Rich Starved the World," *New Statesman,* Apr 17, 2008, www.newstatesman.com/world-affairs/2008/04/food-prices-lynas-biofuels.

18 "The World Food Crisis," editorial, *New York Times,* Apr 10, 2008, www.nytimes.com/2008/04/10/opinion/10thu1.html.

19 Mary Reardon, "Common Questions about ERS Subject Areas," U.S. Department of Agriculture, Economic Research Service, Aug 31, 2010, www.ers.usda.gov/AboutERS/FAQs.htm#hunger.

20 L. Reijnders and S. Soret, "Quantification of the Environmental Impact of Dif-

ferent Dietary Choices," *American Journal of Clinical Nutrition* 78(2003): 664S–668S.

21 Nelleman et al., eds., *The Environmental Food Crisis*, U.N. Environment Programme, 2009, www.grid.no/_res/site/file/publications/Food_crisis_lores.pdf.

22 *2008 Agricultural Statistics Annual,* U.S. Department of Agriculture, National Agricultural Statistics Service, Aug 13, 2009, www.nass.usda.gov/Publications/Ag_Statistics/2008/index.asp.

23 D. Pimentel and M. Pimentel, "Sustainability of Meat-based and Plant-based Diets and the Environment," *American Journal of Clinical Nutrition* 78(2003): 660S–663S.

24 Ibid.

25 L. Reijnders and S. Soret, "Quantification of the Environmental Impact of Different Dietary Choices," *American Journal of Clinical Nutrition* 78(2003): 664S–668S.

26 "China's Rapidly Growing Meat Demand: A Domestic or an International Challenge?," Center for World Food Studies, Dec 2005, www.sow.vu.nl/Brief%20Feed%20for%20China.pdf.

27 F. Zhang, "Beware of Soybean Threat on Chinese Economy," www.EzineArticles.com/?expert=Face_Zhang.

28 *World Agriculture Towards 2030/2050,* U.N. Food and Agriculture Organization, 2006.

CHAPTER 7. HEALTH

1 J. Sabaté, "The Contribution of Vegetarian Diets to Health and Disease: A Paradigm Shift," *American Journal of Clinical Nutition* 78(2003): 502S–507S.

2 H. C. Kung et al., "Deaths: Final Data for 2005," National Vital Statistics Reports, 2008, 56(10), U.S. Centers for Disease Control and Prevention, www.cdc.gov/nchs/data/nvsr/nvsr56/nvsr56_10.pdf.

3 B. Popkin, "Will China's Nutrition Transition Overwhelm Its Health Care System and Slow Economic Growth?," *Health Affairs* 27(2008): 1064–1076.

4 C. Gonzalez, "The European Prospective Investigation into Cancer and Nutrition," *Public Health Nutrition,* 2006, 9(1A): 124–126.

5 "The Nurses' Health Study," www.channing.harvard.edu/nhs/index.php/history/.

6 W. L. Beesom et al., "Chronic Disease among Seventh-day Adventists, a Low Risk Group," *Cancer* 64(1989): 570–581.

7 J. S. Goodwin and M. Brodwick, "Diet, Aging and Cancer," *Clinics in Geriatric Medicine* 11(1995): 577–589.

8 K. Youdin and J. Joseph, "A Possible Emerging Role of Phytochemicals in Improving Age-Related Neurological Dysfunctions: A Multiplicity of Effects," *Free Radical Biology and Medicine* 30(2001): 583–594.

9 D. Steinberg, "A Critical Look at the Evidence for the LDL Oxidation in Atherogenesis," *Atherosclerosis* 131(1997): S5–S7.

10 "Number (in Millions) of Civilian, Non-institutionalized Persons with Diagnosed Diabetes, United States, 1980-2008," U.S. Centers for Disease Control and Prevention, Dec 13, 2010, www.cdc.gov/diabetes/statistics/prev/national/figpersons.htm.

11 Alison Field et al., "Impact of Overweight on the Risk of Developing Common Chronic Diseases during a 10-year Period," *Archives of Internal Medicine* 161(2001): 1581–1586.

12 "Obesity and Overweight," World Health Organization, www.who.int/hpr/NPH/docs/gs_obesity.pdf.

13 "Time to Supersize Control Efforts for Obesity," editorial, *Lancet* 370(2007): 1521.

14 "Overweight and Obesity: Data and Statistics," U.S. Centers for Disease Control and Prevention, Oct 8, 2010, www.cdc.gov/obesity/data/index.html.

15 D. Pimentel and M. Pimentel, "Sustainability of Meat-based and Plant-based Diets and the Environment," *American Journal of Clinical Nutrition* 78(2003): 660S–663S.

16 "Dietary Reference Intakes for Energy, Carbohydrate, Fiber, Fatty Acids, Cholesterol, Protein, and Amino Acids," Food and Nutrition Board, American Institute of Medicine, Sep 5, 2002, http://iom.edu/Reports/2002/Dietary-Reference-Intakes-for-Energy-Carbohydrate-Fiber-Fat-Fatty-Acids-Cholesterol-Protein-and-Amino-Acids.aspx.

17 R. Sinha et al., "Meat Intake and Mortality," *Archives of Internal Medicine,* 2009, 169(6): 562–571.

18 J. Genkinger and A. Koushik, "Meat Consumption and Cancer Risk," *PLoS Medicine,* 2007, 4(12): e345.

19 "Cancer Facts and Figures 2010," American Cancer Society, www.cancer.org/acs/groups/content/@epidemiologysurveilance/documents/document/acspc-026238.pdf.

20 Williamson et al., "Red Meat in the Diet," *Nutrition Bulletin,* 30(2005): 323–355; Norat et al., "Meat, Fish, and Colorectal Cancer Risk: The European Prospective Investigation into Cancer and Nutrition," *Journal of the National Cancer Institute* 97(2005): 906–916.

21 J. C. Lunn et al., "The Effect of Haem in Red and Processed Meat on the Endogenous Formation of *N*-nitroso Compounds in the Upper Gastrointestinal Tract," *Carcinogenesis,* 2007, 28(3): 685–690.

22 J. A. Cross, J. Pollock, and S. Bingham, "Haem, Not Protein or Inorganic Iron, Is Responsible for Endogenous Intestinal N-Nitrosation Arising from Red Meat," *Cancer Research* 63(2003): 2358–2360.

23 P. Lutsey, L. Steffen, and S. Stevens, "Dietary Intake and Development of the Metabolic Syndrome," *Circulation* 117(2008): 754–761.

24 A. Ascherio et al., "Prospective Study of Nutritional Factors, Blood Pressure, and Hypertension among U.S. Women," *Hypertension* 27(1996): 1065–1072.

25 L. Steffen et al., "Association of Plant Foods, Dairy Products, and Meat Intakes with 15-year Incidence of Elevated Blood Pressure in Young Black and White Adults," *American Journal of Clinical Nutrition* 82(2005): 1169–1177.

26 F. Hu and W. Willett, "The Relationship between Consumption of Animal Products and Risk of Chronic Diseases: A Critical Review," Harvard School of Public Health, 1998.

27 E. Cho et al., "Pre-menopausal Fat Intake and Risk of Breast Cancer," *Journal of the National Cancer Institute* 95(2003): 1079–1084.

28 E. Giovannucci et al., A Prospective Study of Dietary Fat and Risk of Prostate Cancer," *Journal of the National Cancer Institute* 85(1993): 1571–1579.

29 Roxanne Khamsi, "Mothers' Beefy Diet Linked to Sons' Low Sperm Count," *New Scientist,* Mar 28, 2007, www.newscientist.com/article/dn11479-mothers-beefy-diet-linked-to-sons-low-sperm-count.html; S. H. Swan et al., "Semen Quality of Fertile US Males in Relation to Their Mothers' Beef Consumption during Pregnancy," *Human Reproduction,* 2007, 22(6): 1497–1502.

30 Marian Burros, "More Salmonella Is Reported in Chickens," *New York Times,* Mar 8, 2006, www.nytimes/com/2006/03/08/dining/08well.html.

31 P. Mead et al., "Food-Related Illnesses and Deaths in the United States," *Emerging Infectious Diseases* 5(1999): 607–625.

32 "Quarterly Enforcement Reports," U.S. Department of Agriculture, Food Safety and Inspection Service, www.fsis.usda.gov/Regulations_&_Policies/QER_Q1_FY2008/index.asp.

33 Kent Garber, "Beef Recall Latest in a Bad Year," *U.S. News and World Report,* Feb 20, 2008, www.usnews.com/news/national/articles/2008/02/20/beef-recall-latest-in-a-bad-year.html.

34 Tiffany Hsu, "Azusa Food Firm Recalls Frozen Beef," *Los Angeles Times,* Aug 8, 2008, http://articles.latimes.com/2008/aug/08/business/fi-beef8.

35 "Dirty Birds: Even Premium Chickens Harbor Dangerous Bacteria," *Consumer Reports,* Jan 2007, 20–23.

36 David Heber and Susan Bowerman, "Applying Science to Changing Dietary Patterns," *Journal of Nutrition* 131(2001): 3078S–3081S.

37 S. Ramassamy, "Emerging Role of Polyphenolic Compounds in the Treatment of Neuro-degenerative Diseases: A Review of Their Intracellular Targets," *European Journal of Pharmacology,* 2006, 545(1): 51–64; E. Gomez-Pinilla, "Brain Foods: The Effect of Nutrients on Brain Function," *Nature Reviews: Neuroscience* 9(2008): 568–578.

38 C. Holick et al., "Dietary Carotenoids, Serum Beta Carotene, Retinal and Risk of Lung Cancer in the Alpha-tocopherol, Beta-carotene Cohort Study," *American Journal of Epidemiology* 156(2002): 536–547.

39 S. Osganian et al., "Dietary Carotenoids and Risk of Coronary Artery Disease in Women," *American Journal of Clinical Nutrition* 77(2003): 1390–1399.

40 E. Giovannucci et al., "Intake of Carotenoids and Retinol in Relation to Risk of Prostate Cancer," *Journal of the National Cancer Institute* 87(1995): 1767–1776.

41 L. L. Marchand et al., "Intake of Flavonoids and Lung Cancer," *Journal of the National Cancer Institute* 92(2000): 154–160; D. F. Birt, S. Hendrich, and W. Wang, "Dietary Agents in Cancer Prevention: Flavonoids and Isoflavonoids," *Pharmacology and Therapeutics,* 2001, 90(2-3): 157–177.

42 C. Yang et al., "Tea Polyphenols Inhibit Cell Hyperproliferation," *Experimental*

Lung Research 24(1998): 629–639; S. Balasubramanian and S. Govindasamy, "Inhibitory Effect of Dietary Flavonol," *Carcinogenesis* 17(1996): 877–879; J. Guo et al., "Dietary Soy Isoflavones and Estrone Protect Wild-type Mice from Carcinogen-induced Cancer," *Journal of Nutrition* 134(2004): 179–182.

43 M. Hertog et al., "Antioxidant Flavonols and Coronary Heart Disease Risk," *Lancet,* 1997, 349(9053): 599; P. Knekt et al., "Flavonoid Intake and Risk of Chronic Diseases," *American Journal of Clinical Nutrition,* 2002, 76(3): 560–568.

44 J. Guo et al., "Dietary Soy Isoflavones Protect Wild-type Mice from Carcinogen-induced Colon Cancer," *Journal of Nutrition,* 2004, 134(1): 179–182; S. Gupta et al., "Inhibition of Prostate Carcinogenesis in Mice by Oral Infusion of Green Tea Polyphenols," *Proceedings of the National Academy of Sciences USA,* 2001, 98(18): 10350–10355.

45 H. Akiyama et al., "Inflammation and Alzheimer's Disease," *Neurobiology of Aging,* 2000, 21(3): 383–421.

46 L. Coussens and Z. Werb, "Inflammation and Cancer," *Nature,* 2002, 420(6917): 860–867.

47 Q. Dai et al., "Fruit and Vegetable Juices and Alzheimer's Disease," *American Journal of Medicine* 119(2006): 751–759.

48 G. McDougall and D. Steward, "The Inhibitory Effects of Berry Polyphenols on Digestive Enzymes," *Biofactors* 23(2005): 189–195.

49 P. Ninfali and M. Bacchiocca, "Polyphenols and Antioxidant Capacity of Vegetables under Fresh and Frozen Conditions," *Journal of Agricultural and Food Chemistry* 51(2003): 2222–2226; V. Bohm et al., "Improving the Nutritional Quality of Microwaved, Vacuum-dried Strawberries," *Food Science and Technology International* 12(2006): 67–75; A. Chaovanalikit and R. Wrolstad, "Total Anthocyanins and Total Phenolics of Fresh and Processed Cherries," *Journal of Food Science* 69(2004): C67–C72.

50 R. Williams et al., "Flavonoids: Antioxidants or Signaling Molecules," *Free Radical Biology and Medicine* 36(2004): 838–849.

51 M. Morris et al., "Association of Vegetable and Fruit Consumption with Age-related Cognitive Change," *Neurology* 67(2006): 1370–1376.

52 M. Donaldson, "Nutrition and Cancer: A Review of the Evidence for an Anti-cancer Diet," *Nutrition Journal* 3(2004): 19–42.

53 F. Hu, "Plant-based Foods and Prevention of Cardiovascular Disease: An Overview," *American Journal of Clinical Nutrition* 78(2003): 544S–551S.

54 J. Anderson and S. Garner, "Phytoestrogens and Human Function," *Nutrition Today* 32(1998): 232; D. Ingram et al., "Case Control Study of Phytoestrogen and Breast Cancer," *Lancet* 350(1997): 990–994; S. Barnes, "Evolution of the Health Benefits of Soy Isoflavones," *Proceedings of the Society for Experimental Biology and Medicine* 217(1998): 386–392.

55 I. Yan et al., "Dietary Flaxseed Supplementation and Experimental Metastasis of Melanoma Cells in Mice," *Cancer Letter* 124(1998): 181–186.

56 W. Demark-Wahnefried et al., "Pilot Study of Dietary Fat Restriction and Flaxseed Supplementation in Men with Prostate Cancer before Surgery," *Urology* 58(2001): 47–52.

57 L. S. Bazzano et al., "Legume Consumption and Risk of Coronary Heart Disease in U.S. Men and Women," *Archives of Internal Medicine* 161(2001): 2573–2578; B. J. Venn and J. I. Mann, "Cereal Grains, Legumes and Diabetes," *European Journal of Clinical Nutrition* 58(2004): 1443–1461; J. D. Potter, "Nutrition and Colorectal Cancer," *Cancer Causes and Control* 7(1996): 127–146; G. R. Howe et al., "Dietary Intake of Fiber and Decreased Risk of Cancers of the Colon and Rectum" *Journal of the National Cancer Institute* 84(1992): 1887–96.

58 P. Slavin, "Why Whole Grains Are Protective: Biological Mechanisms," *Proceedings of the Nutrition Society*, 2003, 62(1): 129–134.

59 Joan Sabaté, "Nut Consumption and Body Weight," *American Journal of Clinical Nutrition* 78(2003), 647S–650S.

60 F. Hu et al., "Frequent Nut Consumption and Risk of Coronary Heart Disease in Women: Prospective Cohort Study," *British Medical Journal* 317(1998): 1341–1345.

61 C. Albert et al., "Nut Consumption and Decreased Risk of Sudden Cardiac Death in the Physicians' Health Study," *Archives of Internal Medicine* 162(2002): 1382–1387.

62 R. Jiang et al., "Nut and Peanut Butter Consumption and Risk of Type 2 Diabetes in Women," *Journal of the American Medical Association* 288(2002): 2554–2560.

63 J. Lampe, "Spicing up a Vegetarian Diet: Chemoprotective Effects of Phytochemicals," *American Journal of Clinical Nutrition* 78(2003): 579S–583S.

64 "How Spicy Foods Can Kill Cancer," BBC News, Jan 9, 2007, www.cayennepepper.info/Spicy%20Foods%20Kill%20Cancer.pdf.

65 "Curry Spice Kills Cancer Cells," BBC News, Oct 28, 2009, http://news.bbc.co.uk/2/hi/8328377.stm.

66 "Popular Curry Spice Is a Brain Booster," *New Scientist,* Aug 4, 2006, www.newscientist.com/article/mg19125635.500-popular-curry-spice-is-a-brain-booster.html.

67 E. Gomez-Pinilla, "Brain Foods: The Effect of Nutrients on Brain Function," *Nature Reviews: Neuroscience* 9(2008): 568–578.

68 "Popular Curry Spice is a Brain Booster," *New Scientist,* Aug 4, 2008, www.newscientist.com/article/mg19125635.500-popular-curry-spice-is-a-brain-booster.html.

69 J. Lampe, "Spicing up a Vegetarian Diet: Chemoprotective Effects of Phytochemicals," *American Journal of Clinical Nutrition* 78(2003): 579S–583S.

70 P. Kris-Etherton et al., "Summary of the Scientific Conference on Dietary Fatty Acids and Cardiovascular Health," *Circulation* 103(2001): 1034–1039; R. Marchioli et al., "Early Protection against Sudden Death by n-3 Polyunsaturated Fatty Acid after Myocardial Infarction," *Circulation* 105(2002): 1897–1903.

71 E. Theodoratou et al., "Dietary Fatty Acids and Colorectal Cancer: A Case Control Study," *American Journal of Epidemiology* 166(2007): 181–195.

72 M. Morris et al., "Consumption of Fish and n-3 Fatty Acids and Risk of Incident Alzheimer Disease," *Archives of Neurology* 60(2003): 940–946.

73 P. Angerer and C. von Schacky, "N-3 Polyunsaturated Fatty Acids and the

Cardiovascular System," *Current Nutrition and Metabolic Care* 3(2000): 439–545.

74 T. van Vilet and M. B. Katan, "Lower Ratio of n-3 to n-6 Fatty Acids in Cultured Than in Wild Fish," *American Journal of Clinical Nutrition* 51(1990): 1–2.

75 Z. Wen and F. Chen, "Heterotrophic Production of Eicosapentaenoic Acid by Microalgae," *Biotechnology Advances* 21(2003): 273–294.

76 "Position of the American Dietetic Association: Vegetarian Diets," *Journal of the American Dietetic Association* 97(1997): 1317–1321.

77 B. Davis and P. Kris-Etherton, "Achieving Optimal Essential Fatty Acid Status in Vegetarians: Current Knowledge and Practical Implications," *American Journal of Clinical Nutrition* 78(2003): 640S–746S.

78 K. Janelle and S. Barr, "Nutrient Intakes and Eating Behavior Scores of Vegetarian and Nonvegetarian Women," *Journal of the American Dietetic Association* 95(1995): 180–188.

79 L. Kushi and E. Giovannucci, "Dietary Fat and Cancer," *American Journal of Medicine* 110(2002): 63S–70S.

80 "Which are Worse: Calories from Carbs or Fat?," *Time*, Jul 15, 2008, www.time.com/time/health/article/0,8599,1822118,00.html.

81 S. Rajaram and J. Sabaté, "Health Benefits of a Vegetarian Diet," *Nutrition* 16(2000): 531–533.

82 W. Roberts, "Preventing and Arresting Coronary Atherosclerosis," *American Heart Journal* 130(1995): 580–600.

83 M. Donaldson, "Nutrition and Cancer: A Review of the Evidence for an Anti-Cancer Diet," *Nutrition Journal* 3(2004): 19–42.

84 A. McMichael et al., "Food, Livestock Production, Energy, Climate Change, and Health," *Lancet* 370(2007): 1253–1263.

85 E. Gomez-Pinilla, "Brain Foods: The Effect of Nutrients on Brain Function," *Nature Reviews: Neuroscience* 9(2008): 568–578.

86 J. Hunt, "Bioavailability of Iron, Zinc, and other Trace Minerals from Vegetarian Diets," *American Journal of Clinical Nutrition* 78(2003): 633S–639S.

87 N. Z. Unlu et al., "Carotenoid Absorption from Salad and Salsa by Humans Is Enhanced by the Addition of Avocado or Avocado Oil," *Journal of Nutrition* 135(2006): 431–436.

88 G. Omenn et al., "Effects of a Combination of Beta Carotene and Vitamin A on Lung Cancer and Cardiovascular Disease," *New England Journal of Medicine* 334(1996): 1150–1155.

89 Lloyd de Vries, "Vitamin E May Shorten Life," CBS News Healthwatch, Nov 10, 2004, www.cbsnews.com/stories/2004/11/10/health/webmd/main654887.shtml.

90 E. Haddad and J. Tanzman, "What Do Vegetarians in the United States Eat?," *American Journal of Clinical Nutrition* 78(2003): 626S–632S.

91 J. Potter, "Food and Phytochemicals, Magic Bullets and Measurement Error: A Commentary," *American Journal of Epidemiology* 144(1997): 1026–1027.

92 "Position of the American Dietetic Association: Vegetarian Diets," *Journal of the American Dietetic Association* 97(1997): 1317–1321.

93 C. Leitzmann, "Vegetarian Diets: What Are the Advantages?," *Forum of Nutrition* 57(2005): 147–156.

CHAPTER 8. DAIRY

1 C. Delgado et al., "Food, Agriculture and the Environment," *Discussion Paper* 28, International Food Policy Research Institute, 1999.

2 I. Oransky, "What's in Your Milk?," *The Scientist*, Feb 2007, 35–40.

3 "Facts About Pollution from Livestock Farms," National Resources Defense Council, www.nrdc.org/water/pollution/farms.asp.

4 Michael Hutjens, "Strategies for Feeding Fat to Dairy Cattle," University of Illinois Extension, Aug 5, 1998, www.livestocktrail.uiuc.edu/dairynet/paperDisplay.cfm?ContentID=246.

5 H. Steinfeld et al., *Livestock's Long Shadow: Environmental Issues and Options,* U.N. Food and Agriculture Organization, 2006, 99.

6 J. Owen, "California Cows Fail Latest Emission Test," *National Geographic News,* Aug 16, 2005, news.nationalgeographic.com/news/2005/08/0816_050816_cowpollution.html.

7 D. Pimentel and M. Pimentel, "Energy Use in Livestock Production," in *Food, Energy, and Society* (Niwot: University Press of Colorado, 1996), 79.

8 Dan Waldner and Michael Looper, "Water for Dairy Cattle," Oklahoma Cooperative Extension Dairy Service ANSI-4275, http://pods.dasnr.okstate.edu/docushare/dsweb/Get/Document-2038/ANSI-4275web.pdf.

9 H. Steinfeld et al., *Livestock's Long Shadow: Environmental Issues and Options*, U.N. Food and Agriculture Organization, 2006, 129.

10 Imke J. M. de Boer, "Environmental Impact Assessment of Conventional and Organic Milk Production," *Livestock Production Science,* Mar 2003, 80(1-2): 69–77.

11 H. Kalkwarf et al., "Milk Intake during Childhood and Adolescence, Adult Bone Density, and Osteoporotic Fractures in U.S. Women," *American Journal of Clinical Nutrition* 77(2003): 257–265.

12 J. Travis, "Got Milk? Dairy Protein Provides Bone-forming Boost," *Science News,* Jun 5, 2004, http://findarticles.com/p/articles/mi_m1200/is_23_165/ai_n6110426/?tag=content;col1.

13 C. M. Weaver, "Should Dairy Be Recommended as Part of a Healthy Vegetarian Diet? Point," *American Journal of Clinical Nutrition* 89(2003): 1634S–1637S.

14 I. Oransky, "What's in Your Milk?," *The Scientist*, Feb 2007, 35–40.

15 D. Sellmeyer et al., "A High Ratio of Dietary Animal to Vegetable Protein Increases the Rate of Bone Loss and the Risk of Fracture in Postmenopausal Women," *American Journal of Clinical Nutrition* 73(2001): 118–122.

16 R. Honkanen et al., "Lactose Intolerance Associated with Fractures of Weight-bearing Bones in Finnish Women Aged 38-57 Years," *Bone* 21(1997): 473–477.

17 P. Appleby et al., "Comparative Fracture Risks in Vegetarians and Non-vegetarians in EPIC-Oxford," *European Journal of Clinical Nutrition* 61(2007): 1400–1406.

18 M. Pereira et al., "Dairy Consumption, Obesity, and the Insulin Resistance Syndrome in Young Adults: The CARDIA Study," *Journal of the American Medical Association* 287(2002): 2081–2089.

CHAPTER 9. SUFFERING

1 "Planning for Breakfast," *Nature*, Feb 22, 2007.
2 M. Dawkins, "Through Animal Eyes: What Behaviour Tells Us," *Applied Animal Behaviour Science* 100(2006): 4–10.
3 T. C. Danbury, et al., "Self-selection of the Analgesic Drug Carprofen by Lame Broiler Chickens," *Veterinary Research* 14(2000): 307–311.
4 C. Riddell, "Developmental, Metabolic, and Miscellaneous Disorders," in *Diseases of Poultry*, 9th ed., ed. Y. M. Saif (Ames: Iowa State University Press, 1991), 831–832.
5 Rick Weiss, "Techno Turkeys," *Washington Post,* Nov 12, 1997.
6 D. Roeber et al., "National Cow and Beef Bull Quality Audit 1999: A Survey of Producer-related Defects in Market Cows and Bulls," *Journal of Animal Science* 79(2001): 658–665.
7 C. Tucker and D. Weary, "Tail Docking in Dairy Cattle," *Animal Welfare Information Center Bulletin,* Winter 2001–Spring 2002, 11(3–4).
8 T. N. Nagaraja and M. M. Chengappa, "Liver Abscesses in Feedlot Cattle: A Review," *Journal of Animal Science* 76(1998): 287–298.
9 D. L. Roeber et al., "Incidence of Injection-site Lesions in Beef Top Sirloin Butts," *Journal of Animal Science* 79(2001): 2615–2618.
10 M. J. Gentle et al., "Behavioural Evidence for Persistent Pain Following Partial Beak Amputation in Chickens," *Applied Animal Behaviour Science* 27(1990): 149–157.
11 Joby Warrick, "'They Die Piece by Piece,'" *Washington Post,* Apr 10, 2001, A01.
12 C. LeDuff, "At the Slaughterhouse, Some Things Never Die: Who Kills, Who Cuts and Who Bosses Can Depend on Race," *New York Times,* Jun 16, 2000, www.nytimes.com/2000/06/16/us/slaughterhouse-some-things-never-die-who-kills-who-cuts-who-bosses-can-depend.html.
13 Robert C. Byrd, "Speaking Up for Animals," *New York State Humane Association Newsletter,* Jul 9, 2001, www.nyshumane.org/articles/speechSenatorByrd.htm.
14 Paul Moran, "Death of a Thoroughbred: Unforgettable Barbaro Gone," *Newsday*, Jan 29, 2007, www.newsday.com/sports/death-of-a-thoroughbred-unforgettable-barbaro-gone-1.554838.

CHAPTER 10. CONSEQUENCES

1 Ovid, *Metamorphoses*, trans. Rolfe Humphries (Bloomington: Indiana University Press, 1960), 268.
2 D. Dombrowski, *The Philosophy of Vegetarianism* (Amherst: University of Massachusetts Press, 1984), 6.

3 Donald H. Reiman and Neil Fraistat, eds., *The Complete Poetry of Percy Bysshe Shelley*, vol. 2 (Baltimore: Johns Hopkins University Press, 2004), 229.

4 Tom Regan, "Utilitarianism, Vegetarianism, and Animal Rights," *Philosophy and Public Affairs* 9 (1980): 305–324.

5 Peter Singer, *Animal Liberation: A New Ethics for Our Treatment of Animals* (New York: New York Review / Random House, 1975).

6 R. Robinson-O'Brien et al., "Adolescent and Young Adult Vegetarians: Better Dietary Intake and Weight Outcome but Increased Risk of Disordered Eating Behaviors," *Journal of the American Dietetic Association,* 2009, 109(4): 648–655.

7 J. Legge and Faxian, *A Record of Buddhistic Kingdoms* (Gloucester, UK: Dodo Press, 2007), 42.

8 "India a Country of Vegetarians? Think Again," *India Abroad,* Oct 17, 2006, www.rediff.com/news/2006/oct/17food.htm.

9 A. Y. Hoekstra and A. K. Chapagain, "Water Footprints of Nations: Water Use by People as a Function of Their Consumption Pattern," *Water Resource Management* (2007) 21: 35–48.

10 Mader et al., "Soil Fertility and Biodiversity in Organic Farming," *Science* 296(2002), 1694–1697.

11 D. Pimentel et al., "Environmental, Energetic, and Economic Comparisons of Organic and Conventional Farming Systems," *BioScience,* Jul 2005, 55(2): 573–582.

12 P. Maeder et al., "Soil Fertility and Biodiversity in Organic Farming," *Science* 296(2002): 1694–1697.

13 A. G. Williams, E. Audsley, and D. L. Sanders, "Energy and Environmental Burden of Organic and Non-organic Agriculture and Horticulture," *Aspects of Applied Ecology* 79(2006): 19–23.

14 Imke J. M. de Boer, "Environmental Impact Assessment of Conventional and Organic Milk Production," *Livestock Production Science,* Mar 2003, 80(1-2), 69–77.

15 "The Organic Myth: Pastoral Ideas Are Getting Trampled as Organic Food Goes Mass Market," *Business Week,* Oct 16, 2006, www.businessweek.com/magazine/content/06_42/b4005001.htm.

16 David Kaimowitz et al., "Hamburger Connection Fuels Amazon Destruction," Center for International Foresty Research, 2009, www.cifor.cgiar.org.

17 W. J. Craig, "Health Effects of Vegan Diets," *American Journal of Clinical Nutrition* 89(2009): 1627S–1633S.

18 *Reducing Poverty and Hunger,* U.N. Food and Agriculture Organization, 2002.

19 Lester Brown, "Could Food Shortages Bring Down Civilizations?," *Scientific American*, May 2009, 50–57.

20 C. Nellemn et al., eds., "The Environmental Food Crisis," U.N. Environment Programme, Feb 2009, www.grid.no/_res/site/file/publications/Food_crisis_lores.pdf.

21 F. Pearce, *When the Rivers Run Dry* (Boston: Beacon Press, 2006), 306.

22 Lester Brown, "Could Food Shortages Bring Down Civilizations?," *Scientific American*, May 2009, 50–57.

23 Lorenzo Cotula et al., "Land Grab or Development Opportunity: Agricultural Investment and International Trade Deals in Africa," U.N. Food and Agriculture Organization, 2009, ftp://ftp.fao.org/docrep/fao/011/ak241e/ak241e00.pdf; Carin Smaller and Howard Mann, "A Thirst for Distant Lands: Foreign Investment in Agricultural Land and Water," International Institute for Sustainable Development, 2009, www.iisd.org/pdf/2009/thirst_for_distant_lands.pdf.

24 "Ocean Temperatures and Sea Level Increases 50 Percent Higher Than Previously Estimated," *Science Daily,* June 19, 2008, www.sciencedaily.com/releases/2008/06/080618143301.htm.

25 C. Mann, "Future Foods: Crop Scientists Seek a New Revolution," *Science* 283(1999): 310–314.

26 "Report of the World Commission on Environment and Development: Our Common Future," U.N. Documents, www.un-documents.net/wced-ocf.htm.

27 Chris Hails, ed., *Living Planet Report 2008,* World Wildlife Fund, Zoological Society of London, and Global Footprint Network, 2, http://assets.panda.org/downloads/living_planet_report_2008.pdf.

28 *Global Environmental Outlook Year Book 4,* U.N. Environment Programme, 2007, 92, www.unep.org/geo/geo4.asp.

29 S. B. Eaton and M. Konner, "Paleolithic Nutrition: A Consideration of Its Nature and Current Implications," *New England Jounal of Medicine* 312(1985): 283–289.

30 "Food Nutrition and the Prevention of Cancer: A Global Perspective," American Institute for Cancer Research, www.aicr.org/site/PageServer?pagenameEQresULreportULsummary.

31 K. Lock et al., "The Global Burden of Disease Attributable to the Consumption of Fruits and Vegetables: Implications for a Global Strategy on Diet," *Bulletin of the World Health Organization* 83(2005): 100–108.

EPILOGUE

1 W. P. T. James et al., "Socioeconomic Determinants of Health: The Contribution of Nutrition to Inequalities in Health," *British Medical Journal,* May 24, 1997, 314(7993): 1545.

2 Kathleen Meister, "Vegetarianism," American Council on Science and Health, Jul 1997, www.asch.org/docLib/20040402(US)Vegetarianism1997.pdf.

INDEX